PHYTOMEDICINE: HEALING HERBS

Healing Herbs

Ven Hari

INDIA • SINGAPORE • MALAYSIA

Notion Press

No.8, 3rd Cross Street,
CIT Colony, Mylapore,
Chennai, Tamil Nadu – 600004

First Published by Notion Press 2020
Copyright © Ven Hari 2020
All Rights Reserved.

ISBN 978-1-63633-717-3

This book has been published with all efforts taken to make the material error-free after the consent of the author. However, the author and the publisher do not assume and hereby disclaim any liability to any party for any loss, damage, or disruption caused by errors or omissions, whether such errors or omissions result from negligence, accident, or any other cause.

While every effort has been made to avoid any mistake or omission, this publication is being sold on the condition and understanding that neither the author nor the publishers or printers would be liable in any manner to any person by reason of any mistake or omission in this publication or for any action taken or omitted to be taken or advice rendered or accepted on the basis of this work. For any defect in printing or binding the publishers will be liable only to replace the defective copy by another copy of this work then available.

Dedication

I dedicate this book to my loving parents who supported me always and the world of plants for supplying food, shelter, clothes, medicine, and the oxygen we breathe and to the great herbalists, teachers of plant biology and all biomedical scientists and practitioners of medicine.

Contents

Appreciation 7
Select Sources for This Book 9

CHAPTER 1	Introduction 13	
CHAPTER 2	History of Phytomedicine 29	
CHAPTER 3	Ethno-Pharmacy 41	
CHAPTER 4	Human Genetics and Genetic Diseases 101	
CHAPTER 5	Cancer, Herbs, Chemoprevention 115	
CHAPTER 6	Herbals for Cardiovascular System 151	
CHAPTER 7	Herbals for Digestive System, Diabetes 165	
CHAPTER 8	Genito-Urinary System and Plants 193	
CHAPTER 9	Herbals for Dermal (Skin) Health 211	
CHAPTER 10	Herbals for ENT and Respiratory Systems 223	
CHAPTER 11	Nervous System, Musculoskeletal, Inflammation, Pain 237	
CHAPTER 12	Anti-Parasitic and Anti-Microbial Herbs 261	
CHAPTER 13	Longevity, Ageing, and Herbals 269	
CHAPTER 14	Alphabetical List of Medicinal Herbs and Phytochemicals 283	

Glossary 335

Appreciation

I would like to acknowledge my indebtedness to my late paternal Grand Father, a retired Forester of the Forest Department of the Government of Travancore now a part of the state of Kerala in India. He was an amateur herbalist who had a great amount of knowledge about medicinal plants that grew in the Western Ghats forests of India and had a good understanding of the basics of Ayurveda. I used to go with him when I was in grade school in his digging operations for finding herbal plants and their roots, and underground stems used in homemade herbal preparations for use by family members. Subsequently, my interest in plant taxonomy and plants in general was stimulated by the excellent lectures by my Professors in India. Although, I subsequently moved on to the study of Virology, Genetics, Cell and Molecular Biology in Adelaide, Australia and subsequently in the USA after immigrating to the United states, I also taught University level courses in Plant development and Biology including lectures on medicinal plants. Being a member of the faculty in Biology, I took part in programs in cancer biology and Virology and Michigan Cancer Institute Laboratories (Karmanos Cancer Inst.) also. After I retired from the University, I kept my interest in the study of plants, plant diseases and medicinal plants. The information on medicinal plants and herbal medical systems presented here is from my own lecture notes and from various books, journals, libraries, and internet sources. All reference sources have been acknowledged.

I am greatly indebted to Professors: K.Rangaswamy Ayyangar (Late) of Annamalai University and Late Prof K.R.Venkatasubban, Prof.Damodaran Thampan and Prof.David Giles Lal of Madras Christian College for their inspiring lectures on Plant taxonomy, anatomy and fungal taxonomy and life cycles and to Prof. T.S. Sadsivan and colleagues of the Center for Advances studies in Botany, Madras University. I also want to acknowledge my mentors and colleagues at the University of Adelaide, University of Arizona and University of California, Berkeley, USA during the formative years of my academic career. This book is the result of the encouragement of my family and friends including some physicians who are interested in alternative medicine.

I want to thank my wife Radha who encouraged me to write this book and my son Raj who in his role as a physician shared his knowledge of medicine. I would also like to express my thanks to all my botanical and medical scientist friends for going over the various chapters and their comments.

Select Sources for This Book

Internet Web sources, text and reference books, original journals, Botanical garden sites and reviews and my personal lecture notes for information for writing and formatting this book. Specific original articles from different journals are cited at the end of each chapter as proper.

Internet-Web

https://forestrypedia.com/engler-and-parantls-system-of-classification/

https://en.wikipedia.org/wiki/Engler_system

https://en.wikipedia.org/wiki/Bentham_%26_Hooker_system

http://www.ville-ge.ch/cjb/herbier_historique_intro.php

https://onlinelibrary.wiley.com/journal/10960031

http://www.theplantlist.org

http://www.motherherbs.com

https://www.mskcc.org/cancer-care/diagnosis-treatment/symptom-management/integrative-medicine/herbs

https://eol.org

https://wcsp.science.kew.org

https://www.kew.org/science

https://www.gbif.org

https://species.wikimedia.org/wiki

http://www.tropicos.org/Home.aspx

http://sweetgum.nybg.org/science

https://www.ncbi.nlm.nih.gov

http://www.catalogueoflife.org

https://plants.jstor.org

https://www.biodiversitylibrary.org

http://www.ville-ge.ch/musinfo/bd/cjb/africa/index.php?langue=an

https://www.bartleby.com/107/(Gray's Anatomy online)

http://www.plantsoftheworldonline.org

https://species.wikimedia.org/wiki/Main_Page

https://www.cbd.int

https://www.anbg.gov.au/apni

http://flora.huh.harvard.edu/china

http://ayush.gov.in

https://plants.jstor.org

http://sweetgum.nybg.org/science

https://ifas.ufl.edu

https://www.cabi.org

https://collections.nlm.nih.gov

https://en.wikipedia.org/wiki/List_of_plants_used_in_herbalism

Book and Journal References

Benzie, I.F.F. and Wachtel-Galor, S. (Eds) (2011): Herbal Medicine, Clinical and Biomolecular aspects, CRC Press, Boca Raton, Florida, USA.

Cohen, Jonathan, William G Powderly, et al. | (2016): Infectious Diseases, 2-Volume Set. 4th edition, Elsevier.

Fauchi, A.S., Braunwald,E, E., Esselbacher,K.J, K.J., Wilson, J.D., Martin, J.B.Kasper, DL., Hauser, S.L, Longo, D.L., (1998): Harrison's Principles of Internal Medicine, 14th edition, McGraw-Hill, New York, pp 2569.

Gray, H (41-st edition, 2015): Gray's anatomy.

Haj.P. Lebrun & A. L. Stork (1991-2015): Enumération des plantes à fleurs d'Afrique tropicale et Tropical African Flowering Plants: Ecology and Distribution, vol. 1-7. Conservatoire et Jardin botaniques de la Ville de Genève.

Hari, Ven (2019): Plant Foods for Nutritional Good Health., ISBN 978-1-68466-906-6 Notion Press, India, pp379.

Nelson, D.S., Cox, M.M. (2017): Lehninger Principles of Biochemistry 7th Edition, ISBN-13: 978-1319125882, ISBN-10: 1319125883, MacMillan Publishers.

CHAPTER 1

Introduction

Phytomedicine is defined as Herbal-based traditional medical practice that uses various plant materials in modalities considered both preventive and therapeutic. Thus, phytomedicine popularly referred as herbal medicine or medicinal plants includes popular folk medicine as practiced in different countries and ethnic groups as well as more organized medicinals based on ancient texts such as traditional Chinese medicine (TCM), Ayurveda, Unani and Siddha medicines of Indian subcontinent and many parts of Asia, Sowa-Rigpa medicine practitioners of Tibet, Greek-Roman treatises, Arab-Persian and Amazonian-South American treatises on medicine as well as some medicinals described in Homeopathy. Collectively, Phytomedicine or herbal medicine forms a substantial part of Alternative holistic medicine including health supplements.

In this book, the term Herbal is used to define all plants and not restricted to those defined as herbs by botanists. The role of plants in supporting good health through proper nutrition and use of plant products in treating disease have been a significant part of human history. Humans realized the use of various wild relatives of plants such as rice, wheat, barley, oats, and other cereals as well as many legumes, oil plants and tubers as sources of food supplementing the flesh of animals for sustenance. In addition, humans realized the value of plants as sources for treating diseases of humans and animals. As civilization advanced, many of these wild plants, food and medicinal plants were domesticated and soon evolved into different species and varieties of food/nutritional and medicinal plants.

The earliest drugs for treatment of various diseases of man and animals came from plants. Many drugs continue to be discovered based on a plant origin. Thus, until the 20^{Th} centuries, many physicians were also botanists- some were professionally qualified botanists and yet others were practicing botanists. Many of these early practitioners of medicine were referred to as medicine men, shamans, sorcerers, herb doctor, witch doctor, Hakims, Ayurvedics etc. Today, people are turning to holistic medicine for better health involving the use of nutritional plant products, herbals of different kinds based on the writings of ancient practitioners of plant-based medicines and folk medicine.

The association of plants as a source for treatment of diseases and for the general well-being of humans came from observation of the behavior of animals that indulged in self-medication, by apparent trial and error, through folk lore, and through ancient works in medicine. These ancient sources are: 1.the Shennong documents from China dating back to about 2500 BC, 2. the Egyptian Ebers papyrus ca 1550 BCE. 3. The Charaka Samhita (uncertain date 4thcentury BCE to the 2nd century) and Sushrutha Samhita (variable days ranging from 600 BCE-600 CE).4. The Bheda Samhita- Bower birch-bark collection of seven distinct manuscripts in San-skrit of 4th -6th Century C. E currently held at the Bodleian Library of Oxford University 5. Ashtanga Hridaya Samgraha of Vaghabhatta ca 700 CE. and the Rug-vinischaya or Madadhava Nidana composed by Madhva-kara (700 CE). 6. Sharangadhara Samhita (9-13th century CE). 7. Bhavaprakasha (14th century CE, from India).8. The Greek scientists-philosophers -Hippocrates 460 BCE, Aristotle 384 BCE, Theophrastus. 371–287 BCE, Galen 129-215 CE, De Materia Medica of Dioscorides (AD 40- c. 90), and the canon of Medicine by Avicenna 980-1037 CE. 9. The compendium of Materia medica or Bencao Gangmu compiled during the Ming dynasty (1578 CE).

Added sources of information include the ethnic botanical studies conducted on Australian aborigines, African tribal-folk medicine, rainforest populations in South America and American Indian folk medicine. Further, the ability to extract and test the components of many herbs for drug activity through modern chemical analysis techniques has made it possible to synthesize many drugs that have a plant origin and produce modified or altered therapeutic drugs of plant origin so that the altered drugs were more potent and less toxic to the patient being treated.

Plants are the source of carbohydrates, proteins, sugar, Fats, fiber, minerals, and vitamins. In addition, they contain compounds known as Phytochemicals that are synthesized in plants as adaptive mechanisms to withstand attacks on them by insects, nematodes, parasites and various types of microorganisms as well as a means to withstand stresses and react to environmental impact. These compounds are present in the entire plant or are in specific parts of the plant such as leaves, rhizome, roots, bark, and seeds. Thus, the stimulant/decongestant ephedrine can be obtained from the entire plant *Ephedra sinica* whereas, the anti-malarial Quinine comes mainly from the bark of *Cinchona* sp, the tranquilizer reserpine from roots of the snake root plant (*Rauwolfia serpentina*), the steroid Diosgenin from tubers of *Dioscorea* sp and the Central nervous system stimulant strychnine from the nuts of *Strychnos nux vomica*. Further, the quantities of the active compounds in a plant vary during growth seasons end even the time of day. The phytochemicals, which are known as secondary metabolites, show anti-microbial, anti-oxidative, and cytotoxic properties that often have medicinal value in veterinary and human medicine. These phytochemicals fall into the following chemical classes namely: terpenes, phenolics, glucoseinolates, Betalain, indoles, organo-sulfur compounds, piperines, Chlorophylls, organic acids, and proteinases. In addition, plants such as *Plantago ovata* (*Psyllium*) and many food grains have fibers that are especially important for gastrointestinal and cardiovascular health.

Many drug companies and Health Institutes of various countries have been sending out teams of ethno-botanists and other scientists to various remote regions of the world to identify potential medicinal plants based on tribal and folk lore and animal behavior. The National Cancer Institute (NCI) and the National Center for Complementary and Integrative Health(NCCIH) of the National Institute of Health (NIH) in the USA has established several collaborative programs with Countries in Africa, Australasia, Europe, and South America to look and screen plants for medicinal value particularly for treatment of Cancers, heart disease, viral, bacterial, and parasitic diseases. Similar programs have been initiated by several other countries especially by the Medical Research Council (MRC) of the United Kingdom, organizations in the European Union, the Central Drug research Institute (CDRI) and Council of Scientific and Industrial Research in India (CSIR), The Council of Scientific research organization of Australia (CSIRO) and many similar organizations elsewhere. Many plant-based products that are either already approved or are in various stages of clinical trials have been listed and described in the following chapters.

Plant based medicines also known as Herbal medicines, Botanicals and Phytomedicine are becoming extremely popular among people who are concerned about the cost of allopathic medicines, side effects of many medicines and because of the popular belief that phytomedicines are safer to use because of their organic origin. The WHO has estimated that in many third world countries about 80% of medications used in primary care are botanicals. Even in a country like Germany, physicians prescribe 600-700 plant-based medicines to their patients. This market is spurred by ageing populations and the belief by many that Natural remedies are safer than the scientifically evaluated allopathic medicines that are marketed by the major drug companies. However, one major concern is that many of the plant-based medicines have not been evaluated for safety and even for efficacy. As a matter of fact, studies conducted by the Alternative medicine group of the National Institute of Health USA (NIH)and by the Medical Council of United Kingdom (MRC) show that many of the herbal medicines contain harmful levels of heavy metals such as lead, mercury, cadmium, synthetic chemicals, fungicides and pesticides. Hence, care should be exerted in consuming these herbal preparations. Obviously, more care should be given to set up better quality control in the manufacture and formulation of these herbal preparations. Nevertheless, it is clear that many plants contain metabolic products that have both curative and preventative products that are conducive to good health and that further studies will identify more phytochemicals that will help to improve good health. As of now, there are over 120 important drugs such as Atropine, coumarins which are pre-cursors of anticoagulants like Warfarin, Digitalis alkaloids that are Cardiotonic, Diosgenin which is the progenitor of corticosteroids and birth control pills, the stimulant ephedrine, the anti-malarial Quinine and Artemesinin, anticholinergic, and muscarinic scopolamine, anti-cancer drugs Taxol, Vincristine, Vinblastine and others that have a plant based identity.

Table 1: List of phytochemicals with known mode of action. Some of these is incorporated in FDA list of approved allopathic drugs. Some of these are over the counter drugs (OTC) which are marketed because they follow FDA regulations and are not prescription medications.

Names of drugs were verified with the International drug name database: https://www.drugs.com/international/.

DRUG	THERAPEUTIC USE	PLANT DERIVED FROM
Acetyl digoxin	Cardio tonic	*Digitalis lanata* (Grecian foxglove, woolly foxglove)
Adoniside	Cardio tonic	*Adonis vernalis*, (pheasant's eye, red chamomile)
Aescin	Anti-inflammatory	*Aesculus hippocastanum*, (horse chestnut)
Aesculetin	Anti-dysentery	*Frazinus rhynchophylla*
Agrimophol	Anthelmintic	*Agrimonia eupatoria*
Ajmalicine	circulatory disorders	*Rauvolfia serpentina* (Snakeroot)
Allantoin	Vulnerary wound healing	several plants
Allyl isothiocyanate	Rubefacient	*Brassica nigra* (black mustard)
Anabesine	Skeletal muscle relaxant	*Anabasis sphylla*
Andrographolide	bacillary dysentery/respiratory	*Andrographis paniculata*
Anisodamine	Anticholinergic	*Anisodus tanguticus*
Anisodine	Anticholinergic	*Anisodus tanguticus*
Arecoline	Anthelmintic	*Areca catechu* (betel nut palm)
Artemesinin	Anti-Malarial	*Artemisia annua*
Asiaticoside	Wound healing	*Centella asiatica* (gotu cola)
Atropine	Anticholinergic	*Atropa belladonna* (deadly nightshade)
Benzyl benzoate	Scabicide	several plants
Berberine	bacillary dysentery	*Berberis vulgaris* (common barberry)
Bergenin	Antitussive	*Ardisia japonica* (Marlberry, Hakuokan)
Betulinic acid	Anti-cancerous	*Betula Alba* (common birch)
Borneol	Antipyretic, analgesic,	several plants
Bromelain	Anti-inflammatory, proteolytic	*Ananas comosus* (pineapple)

Continued...

Phytomedicine: Healing Herbs

Caffeine	CNS stimulant	*Camellia sinensis* (tea), *Coffea arabica* (coffee), *Theobroma cacao*, (cocoa)
Camphor	Rubefacient	*Cinnamomum camphora* (camphor tree)
Campothesin	Anti-cancerous	*Campotheca acuminata*
Cannabinoids	Chronic Pain, Nausea, Vomiting Chemotherapy side effects	*Cannabis sativa* (Hemp, marijuana)
Catechin	Hemostatic	*Potentilla fragarioides*
Chymopapain	Proteolytic, mucolytic	*Carica papaya* (Papaya fruit latex)
Cissampeline	Skeletal muscle relaxant	*Cissampelos pareira* (velvetleaf)
Cocaine	Local anesthetic	*Erythroxylum coca* (coca plant)
Codeine	Analgesic, antitussive	*Papaver somniferum* (poppy latex)
Colchicine amide	antitumor agent	*Colchicum autumnale* (autumn crocus)
Colchicine	Antitumor, anti-gout	*Colchicum autumnale* (autumn crocus)
Convallatoxin	Cardiotonic	*Convallaria majalis* (lily-of-the-valley)
Coumarin	pre-anti-coagulant/Warfarin	*Dipteryx odorata* (Tonka beans), Also, *Galium odoratum*
Curcumin	Choleretic, multiple action	*Curcuma longa* (Turmeric rhizome)
Cynarin	Choleretic	*Cynara scolymus* (artichoke fruit)
Danthron	Laxative	*Cassia* species
Demecolcine	Antitumor agent	*Colchicum autumnale* (autumn crocus)
Deserpidine	Antihypertensive, tranquilizer	*Rauvolfia canescens*
Deslanoside	Cardiotonic	*Digitalis lanata* (Grecian foxglove, woolly foxglove)
L-Dopa	Anti-parkinsonism	*Mucuna* species (cowage, velvetbean)

Digitalin	Cardiotonic	*Digitalis purpurea* (purple foxglove)
Digitoxin	Cardiotonic	*Digitalis purpurea* (purple foxglove)
Digoxin	Cardiotonic	*Digitalis purpurea* (purple or common foxglove)
Dronabinol	Appetite stimulant	Synthetic THC analog based on Cannabis sativa
Emetine	Amoebicide, emetic	*Cephaelis ipecacuanha* (Ipecac)
Ephedrine	Sympathomimetic, antihistamine	*Ephedra sinica* (Ephedra, ma huang)
Epidiolex	Lennox-Gastaut and Dravet syndrome	Epilepsy-Cannabidiol from Cannabis sativa
Epicatechin	Cardiovascular	Cardio protection (Coca, tea, prune)
Etoposide	Antitumor agent	*Podophyllum peltatum* (Mayapple)
Eucalyptus	Muscle pain, headache	Eucalyptus sp.
Galanthamine	Cholinesterase inhibitor	*Lycoris squamigera* (magic lily, resurrection lily)
Gitalin	Cardiotonic	*Digitalis purpurea* (foxglove)
Glaucarubin	Amoebicide	Simarouba glauca (paradise tree)
Glaucine	Antitussive	*Glaucium flavum* (poppy, sea poppy)
Glasiovine	Antidepressant	*Octea glaziovii*
Glycyrrhizin	Sweetener, Addison's disease	*Glycyrrhiza glabra* (licorice)
Gossypol	Male contraceptive	*Gossypium species* (cotton seeds)
Hemsleyadin	bacillary dysentery	*Hemsleya amabilis*
Hesperidin	Treatment for capillary fragility	*Citrus sp.* (oranges, lemons, lime etc)
Hydrastine	Hemostatic, astringent	*Hydrastis canadensis* (goldenseal)
Hyoscyamine	Anticholinergic	*Hyoscyamus niger* (henbane,)
Irinotecan	Anticancer, antitumor agent	*Camptotheca acuminata*

Continued...

Kaibic acid	Ascaricide	*Digenea simplex* (wireweed)
Kawain	Tranquilizer	*Piper methysticum* (kava kava)
Kheltin	Bronchodilator	*Ammi visaga*
Lanatosides A, B, C	Cardiotonic	*Digitalis lanata* (Grecian foxglove, woolly foxglove)
Lapachol	Anticancer, antitumor	*Tabebuia species* (Trumpet tree)
a-Lobeline	Smoking deterrant, stimulant	*Lobelia inflata* (Indian tobacco)
Lutein	Eye health	Kale, Spinach, Carrots, Pot Mari gold, Nasturtium, Broccoli
Marinol	Appetite stimulant	synthetic based on *Cannabis sativa*
Menthol	Rubefacient	*Mentha species* (Mint)
Methyl salicylate	Rubefacient	*Gaultheria procumbens* (wintergreen)
Minnelide	Anticancer analog of Triptolide	*Tripterygium wilfordii* (Lei Gong Teng)
Monocrotaline	Topical antitumor	*Crotalaria sessiliflora*
Morphine	Analgesic	*Papaver somniferum* (poppy)
Neoandrographolide	Treatment of dysentery	*Andrographis paniculata*
Nicotine	Insecticide	*Nicotiana tabacum* (tobacco)
Nordihydroguaiaretic acid	Antioxidant	*Larrea divaricata* (creosote bush)
Noscapine	Antitussive	*Papaver somniferum* (poppy)
Ouabain	Cardiotonic	*Strophanthus gratus* (ouabain tree)
Pachycarpine	Oxytocic	*Sophora pschycarpa*
Palmatine	Antipyretic, detoxicant	*Coptis japonica* (Chinese goldenthread, Huang-Lia)
Papain	Proteolytic, mucolytic	*Carica papaya* (Papaya)
Papavarine	Smooth muscle relaxant	*Papaver somniferum* (Opium poppy, common poppy)
Phyllodulcin	Sweetener	*Macrophylla* (Bigleaf hydrangea, French hydrangea)
Physostigmine	Cholinesterase inhibitor	*Physostigma venenosum* (Calabar bean)

Picrotoxin	Analeptic, Anamirta	*Cocculus* (Fish berry)
Pilocarpine	Parasympathomimetic	*Pilocarpus jaborandi*, (Indian hemp)
Pinitol	Expectorant	Several plants e.g. Bougainvillea
Piperine	Bioavailability of drugs	*Piper nigrum* (Black peeper), *P. longum* (Long pepper)
Podophyllotoxin	Antitumor, anticancer agent	*Podophyllum peltatum* (Mayapple)
Protoveratrines A, B	Antihypertensives	*Veratrum album* (white false hellebore)
Protodioscin	Testosterone booster	Body building, Sex
Pseudoephedrine	Sympathomimetic	*Ephedra sinica* (Ephedra, Ma huang)
Nor-pseudoephedrine	Sympathomimetic	*Ephedra sinica* (ephedra, Ma huang)
Quinidine	Antiarrhythmic	*Cinchona officianalis* (Quinine tree)
Quinine	Antimalarial, antipyretic	*Cinchona officianalis* (Quinine tree)
Qulsqualic acid	Anthelmintic	*Quisqualis indica* (Rangoon creeper, drunken sailor)
Rescinnamine	Antihypertensive tranquilizer	*Rauvolfia serpentina* (Snakeroot)
Reserpine	Antihypertensive, tranquilizer	*Rauvolfia serpentina* (Snakeroot)
Rhomitoxin	Antihypertensive, tranquilizer	*Rhododendron molle* (Rhododendron)
Rorifone	Antitussive	*Rorippa indica*
Rotenone	Piscicide, Insecticide	*Lonchocarpus nicou*
Rotundine	Analagesic, sedative, tranquilizer	*Stephania sinica*
Rutin	Treatment for capillary fragility	*Citrus species* (Orange, grapefruit, lime, etc)
Salicin	Analgesic	*Salix alba* (White willow)
Sanguinarine	Dental plaque inhibitor	*Sanguinaria canadensis* (Bloodroot)
Santonin	Ascaricide	*Artemisia maritma* (Wormwood)

Continued...

Scillarin a	Cardiotonic	*Urginea maritima* (squill)
Scopolamine	Sedative	*Datura species* (Jimsonweed)
Sennosides A, B	Laxative	*Cassia sp.* (Cinnamon)
Shikimic acid	Source for drug Tamiflu	*Illicium verum* (Star Anise)
Silymarin	Antihepatotoxic	*Silybum marianum* (Milk thistle)
Sparteine	Oxytocic	*Cytisus scoparius* (Scotch broom)
Stevioside	Sweetener Sugar substitute	*Stevia rebaudiana* (Stevia)
Strychnine	CNS stimulant	*Strychnos nux-vomica* (Poison nut tree)
Taxol	Antitumor agent	*Taxus brevifolia* (Pacific yew)
Teniposide	Antitumor agent	*Podophyllum peltatum* (Mayapple)
a-Tetrahydrocannabinol	THC Antiemetic, ocular tension	*Cannabis sativa* (Marijuana)
Tetrahydropalmatine	Analgesic, sedative, tranquilizer	*Corydalis ambigua*
Tetrandrine	Antihypertensive	*Stephania tetrandra*
Theobromine	Diuretic, vasodilator	*Theobroma cacao* (Cocoa)
Theophylline	Diuretic, bronchodilator	*Theobroma cacao*, Thea sinensis (Cocoa, Tea)
Thymol	Topical antifungal	*Thymus vulgaris* (Thyme)
Topotecan	Antitumor, anticancer	*Camptotheca acuminata* (Cancer/Happy tree)
Trichosanthin	Abortifacient	*Trichosanthes kirilowii* (snake gourd)
Tubocurarine	Skeletal-muscle relaxant	*Chondodendron tomentosum* (curare vine)
Valapotriates	Sedative	*Valeriana officinalis* (valerian)
Vasicine	Cerebral stimulant	*Vinca minor* (periwinkle)
Vinblastine	Antitumor, Anti-leukemic	*Catharanthus roseus* (Madagascar periwinkle)
Vincristine	Antitumor, Anti-leukemic	*Catharanthus roseus* (Madagascar periwinkle)
Warfarin	Anti-coagulant 4-hydroxycoumarin	*Melilotus* (Fungus infected sweet clover)

Xanthotoxine	Vitiligo, Leucoderma	*Ammi majus*
Yohimbine	Aphrodisiac	*Pausinystalia johimbe* (yohimbe)
Yuanhuacine	Abortifacient	*Daphne genkwa* (Thymeliaceae - Malvales)
Zeaxanthin	Eye health	Corn, Paprica, Saffron, Wolfberries

Notes about Medical Terms

Abortifacient = facilitates abortions; Analeptic = restorative; Analgesic = pain killer; Anthelmintic = anti-worm (mostly tape worm, nematodes); Antiarrhythmic = prevents rapid heartbeats; Antihepatotoxic = protects against liver toxicity; Antipyretic = reduce fever, Anti-tussive = cough suppressant; Aphrodisiac = stimulates sexual desires; Ascaricide = kills round worms; CNS = Central nervous system; mucolytic = loosens mucus; Cholinesterase inhibitor = Cholinesterase inhibitors are drugs that block the activity of an enzyme in the brain called cholinesterase. Cholinesterase breaks apart the neurotransmitter acetylcholine, which is vital for the transmission of nerve impulses. cholagogues = Stimulants that stimulate flow of bile into duodenum, Choleretic = Stimulates bile from liver, hemostatic = agents that control bleeding, hypoglycemic = low sugar in body, Oxytocic = stimulating contraction of the uterus, Rubefacient = topical application causing redness of skin. Sympathomimetic = drugs mimic or enhance the actions of endogenous catecholamine's of the sympathetic nervous system, Vitiligo = white patches on skin or hair.

About Herbal Plants and Medicine

Plants have many chemicals collectively referred to as phytochemicals with similar and different properties. These chemicals protect the plants against invading organisms, stress, environmental problems and for their own physiology for growth, reproduction, survival, and dispersal of seeds. However, as it turns out many of these plant phytochemicals have both beneficial and harmful effects on human and animals. The traditional herbalist finds these properties, prescribes extracts of these plants, and adds them in specific formulations for treatment. Thus, with very few exceptions these herbal preparations are prescribed in combination with other herbals mainly because herbalists treat a disease as a reflection of multiple disruptions in the body rather as an isolated effect on just one part of the body. For example, an herbalist sees a skin condition such as eczema or psoriasis or acne not as a single symptom resulting from a single cause but as one whose cause lies in the immune system, inflammatory system, skin growth etc and treats it from a holistic approach.

In these forthcoming chapters, while naming the potential of an herb for a health condition, it is realized that the same herb could be useful for treating other health conditions.

Although, herbal products are being marketed and used by many individuals, it must be noted that the use of these products with various often-unsubstantiated claims may be subject to various levels of regulations in various countries. The World health Organization (WHO) has published a document on quality control methods to help countries to consider regulating and applying quality control standards for the marketing of herbal and other medical products. In the USA, herbal remedies are not required to prove efficacy or safety of the products but the Food and Drug Administration (FDA) can impose restrictions on the sale of a product when there is evidence that the product may be harmful. A classic example is Marijuana whose sale is regulated or whose sale for limited usage must be approved by the concerned states based on voter approval. Likewise, the sale of ephedra as a dietary supplement is prohibited by the Food and Drug Administration, USA (FDA).

Disclaimer

The various chapters in this book have pertinent information about History, ethnic biology, descriptions of medicinal plants, phytochemicals as well as herbals of current and or potential use of plants in herbal preparations in treating disease. At the same time, I wish to state that this book is only a compendium of medicinal plants to be used as a source of information for academic purposes, research, product information and development of new health care products and as a point of reference. It is not meant as a recommendation for use of any of the herbal products for conditions specified by herbalists. **A qualified physician is the best person to diagnose and treat any disease.**

Herbal Delivery

Herbal remedies are rarely composed of a single herb. Usually, many different herbal sources from one or more herbs are mixed in different proportions along with carriers for treating different medical conditions. Thus, most herbal formulations known as poly-herbals have extracts from more than one herb. Since individual plant species have many different phytochemicals, it is common to find an herb that is prescribed for more than one condition.

Herbal medicines are supplied and used as powders, liquids, herbal ashes, capsules, tablets, granules, paste, Creams, gels, decoctions and herbal extracts in Oil (Coconut, Sesame, olive, mustard) or aqueous media, Steam distillates and alcoholic herbal tonics. Most modern clinical tests reported in the literature are conducted on single phytochemical components in a plant and as such the results obtained need not reflect on the whole plant products which have multiple phytochemicals and this is especially so with respect to polyherbal formulations.

Endocrine System

The endocrine system is made up of all glands that secrete different types of hormones such as the Adrenal, Pituitary, thyroid, prostaglandins, Estrogen, progesterone, testosterone, insulin, erythropoietin, eicosanoids, leukotrienes, Angiotensin and other hormones which play a major role in homeostasis and overall control of the working of all systems. A number of glands that signal each other in sequence such as the hypothalamic-pituitary-adrenal axis and many other organs that are part of other body systems have secondary endocrine functions, including bone, kidneys, liver, heart and gonads. In view of multiple organ involvement, a separate chapter has not been devoted to this system and information on herbals that affect the endocrines are distributed throughout this book in various chapters.

Plant Taxonomy

Several forms of Plant taxonomy are in use. Although, this author is more familiar with the system of classification as per Bentham and Hooker, as well as that of Engler and Prantl and Hutchinson, I have chosen to use the classification based on the APG IV system (Angiosperm Phylogeny Group) released in 2016. This system is based on DNA molecular sequence analysis data and cladistics methods as applied to biochemical and molecular genetic traits of organisms, as well as to anatomical ones. It is the system most currently followed since it considers relationships based on common evolutionary ancestry. The binomial system (Species Plantarum) of Carl Linnaeus (1707-178) is the basis of naming the plant genera and species as in all other systems of classification.

References

https://www.nlm.nih.gov/about/herbgarden/list.html

https://www.fda.gov/food/dietarysupplements

https://nccih.nih.gov/health/supplements/wiseuse.htm

https://www.opss.org/prohibited-department-defense

https://www.deadiversion.usdoj.gov/schedules

http://www.ahpa.org

https://www.accessdata.fda.gov/scripts/cder/daf

https://www.merriam-webster.com/medical

http://www.historyworld.net/wrldhis/PlainTextHistories.asp?historyid=aa52

https://www.britannica.com/science/history-of-medicine/Traditional-medicine-and-surgery-in-Asia

https://www.britannica.com/science/traditional-Chinese-medicine

https://www.ncbi.nlm.nih.gov/pmc/articles/PMC4379645

http: //tropical.theferns.info

https://www.feedipedia.org

https://www.analogforestry.org/resources/database

http://www.plantsoftheworldonline.org

https://www.ipni.org

https://www.alwaysayurveda.com

The Angiosperm Phylogeny Group, (Chase, M. W., Christenhusz, M. J. M., Fay, M. F., Byng, J. W., Judd, W. S, Soltis, D. E., Mabberley, D. J., Sennikov, A. N., and P. S. Soltis (2016): An update of the Angiosperm Phylogeny Group classification for the orders and families of flowering plants, APG IV. Botanical Journal of the Linnaean Society, Volume 181, Issue 1, 1 May 2016, Pages 1–20.

Bentham, G., and Hooker, J.D. (1862–1883): Genera plantarum ad exemplaria imprimis in herbariis kewensibus servata definita (3 volumes).

Bhattacharyya, B., (2005): Systematic Botany, Harrow: Alpha Science International Ltd. ISBN 9781842652510 Retrieved 29 April 2015.

Davi-Ellen Chabner (1978): Medical Terminology: A Short Course, 7th edition, Elsevier publications.

Hutchinson, John (1973): The families of flowering plants, arranged according to a new system based on their probable phylogeny. 2 Vols (3rd Ed.), Oxford University Press.

Morley, Thomas (1984): "An Index to the Families in Engler and Prantl's "Die Natürlichen Pflanzenfamilien', Annals of the Missouri Botanical Garden, 71 (1): 210–228. JSTOR 2399064.

Naik, V.N. (1984): Taxonomy of Angiosperms New Delhi: Tata McGraw-Hill Education. ISBN 9780074517888.

Roy, H. History of Medicine with Special Reference to India, from web source, http://www.histopathology-india.net/history_of_medicine.htm,

Sushruta Samhita (Sushruta' a Collection) (800-600 B.C.?). Pioneers of plastic surgery Acta Chir Plast., 1984; 26(2): 65-8.

Suśruta: a man of history and science, Int Surg. 1968 Nov; 50(5): 403-7.

CHAPTER 2

History of Phytomedicine

Early History: Knowledge about food and medicinal value of plants date back to prehistoric times as shown by pollen from food and medicinal plants buried in paleolithic graves. This knowledge apparently came from trial and error, through religious beliefs that God created specific plants for helping humans to treat disease that was elaborated in the disease doctrine of Signatures, observation of the self-medication behavior of animals such as chimpanzees, spider monkeys, gorillas, birds and other animals as well as folk lore. The known ancient and middle age written records of speculated dates are:

- The Assyrian King Ashurbanipal who listed over 300 medicinal plants including opium and myrrh.

- The Ebers papyrus from Egypt listing 850 plants: The scroll contains 700 magical formulas and folk remedies meant to cure afflictions ranging from crocodile bite to toenail pain and to rid the house of such pests as flies, rats, and scorpions. It also includes a surprisingly correct description of the circulatory system, noting the existence of blood vessels throughout the body and the heart's function as center of the blood supply.

- The Rig Veda 1500 BCE: Knowledge of the Hymns of Praise", for recitation.

- Atharva Veda 1000 BCE: "Knowledge of the Magic formulas" These ancient texts have information on herbals and medicine.

- Shennong documents from China dating back to about 2500 BC. Shennong Bencao Jing (Shennong Native Herbs Anthology) is considered the first Chinese monograph on pharmacology. It itemized 365 medicinal products.

- The Bencao Gangmu or Pen-tsao Kang-mu a Chinese materia medica work compiled by Li Shizhen during the Ming Dynasty (LI Shizhen,1518~1593) listing 1892 herbs. The basis of hood health is to reach a proper balance between the Human body and the natural world and the components of the natural world.

- The Historia Plantarum of Theophrastus 350 BCE – c. 287 BCE

- Charka and Sushrutha Samhita from India 3rd Century C.E. listing over 700 medicinal plants.

- The Bheda Samhita- Bower manuscripts 4th-6th Century C.E that are birch-bark collection of seven distinct manuscripts in the Sanskrit language written on bark parchment and currently held at the Bodleian Library of Oxford University.

- The Ashtanga Hridaya of Vaghabhatta 6th C.E.

- Madhava Nidana by Madhavakara 7th Century,

- Ratirahasya by Kokkoa in 9th Century.

- Sarangadhara Samhita by Sharangadhara in 14th century,

- Bhavaprakasa by Bhava Mishra 15th century and Ananga Ranga by Kalyanamalla in 16th century, Vaidya rathnavali by Raja Serfoji in 1798-1832.

- Siddha is an ancient Indian traditional treatment system which evolved in South India, and is dated to the times of 3rd millennium BCE, Indus Valley Civilization or earlier

- *The De Materia Medica* of Dioscorides, which is a five-volume encyclopedia of medicinal plants First century C.E.

- The Canon of Medicine by Avicenna Galen from Persia 980-1037 CE.

- The De Historia Stirpium 1572 C.E. of Leonhard Fuchs of Germany.

- The Herbarum Vivae Eicones 1530-1536 CE of Otto Brunfels.

- New Kreuterbuch 1546 C.E. of Hieronymus Bock.

- Dispensatorium 1536 of Valerius Cordus 1515-1544 C.E.

- Herbal 1525 of Richard Bancke from England.

- General History of Plantes 1597 of John Gerard.

- Materia Medica and Species Plantarum of Carl Linnaeus 1753 C.E.

- The Aztec medicinal plants known as the Badianus Herbal or The Libelous de Medicinalibus Indorum Herbis or codex berberini translated from Aztec language by Martin de La Cruz and Juannes Badianus of the Roman Catholic College of Santa Cruz, Peru, discovered in the Vatican library and currently in the National Institute of Anthropology and History in Mexico City. This manuscript is dated to 1520 C.E.

- Unani medicine texts based on Arabic-Persian-Greek medicine.

The Doctrine of Signatures: One of the main bases for the choice of specific plants as having medicinal value is the doctrine of signatures. The doctrine of signatures holds that God designed plants on earth such that they bore specific signs or signatures that would help man to find them for the treatment of diseases. The Chinese and later many European Herbalists believed that human destiny was controlled by planets/stars and that God designed plant morphology (appearance) to reflect their potential use for human welfare. Thus, the shape of a plant or it's organ, the texture of the plant and it's parts, the color of the plant and it's flowers, the juices produced by plant sap, the smell of the plant and the habitat of the plant were all considered to be reflective signs to enable man to identify them for use as medicines. Thus, for example, the heart shape of the leaves of a plant is a signature for potential use for treating heart diseases. Likewise, if the juice of a plant was yellow in color, it was considered useful in treating liver or gall bladder conditions. There is no evidence that this doctrine based on morphology and other external features of the plant are of any great help in finding medically useful plants. Nevertheless, this concept had some role in the classification of plants that was eventually codified on a scientific basis by Linnaeus in 1753 and other plant taxonomists. Indeed, some plants did turnout to have some potential as medically relevant plants and even today, many practitioners of alternative medicine continue to recommend products from plants based on the doctrine of signatures as a treatment for many diseases.

Although, this concept was formalized during the middle ages in Europe, similar concepts existed in other cultures even earlier. The Chinese believed that plant structures correlated with shapes of human organs. As per the yin- Yang concept, diseases of the upper parts of humans were treated with plant parts that were above ground and lower parts of the body were treated by below ground plant parts. Similarly, in India, the sage Agasthya is said to have had the ability to speak to plants and obtain information about their usefulness. In Europe, the writings of Galen A.D. 131-200 gave hints that eventually formed the basis of the doctrine of signatures.

The Swiss Scientist and Philosopher, Paracelsus who lived in the 16th Century 1491-1541 was one of the most famous proponents of the doctrine of signatures and his ideas received great impetus from the Church, which reasoned that the Almighty would have provided signs whereby plants could be identified for the treatment of human diseases. He developed astrological talismans for curing disease. Jakob Bohne (1575-1624) claimed that he had a vision in which he saw signals that proved the relationship between God and Man. He codified his thoughts in the form of a book known as "Signatura Rerun" or Signature of all things in 1621 in which he put a spiritual spin on the doctrine whereby a relationship between macrocosm and microcosm was established. In simple terms, the doctrine essentially meant that the resemblance of plants to various human body parts is a signature from the Creator so that physicians can identify specific plants as sources for treatment of ailments affecting various body parts. As this doctrine developed further, liverwort was suggested for treating liver conditions, and similarly snake wort, for snake bites, lungwort for

History of Phytomedicine | 33

lung diseases, bloodwort for blood conditions, toothwort for treating teeth conditions, wormwood for treating intestinal diseases, Lousewort for treating Lice, Spleenwort for Spleen conditions. Other examples include Plants with yellow flowers or yellow juice for treating jaundice, Maidenhair fern for treating Baldness, Walnut for brain conditions etc.

An added dimension to the doctrine of Signatures was given by the British author Nicholas Culpepper (1616-1654). His books *"the English Physician"* 1652, *Complete Herbal* (1653) and *"Astrological Judgment of Diseases from the Decumbiture of the Sick"* (1655) described many plants that were considered to have medicinal value. He connected Astrology and the Doctrine of signatures in what was popularly called Medical Astrology. Briefly, many people believed that various planets controlled various aspects of life, likewise specific plants were associated with specific planets, and hence whichever disease was being influenced by the planets would be cured by the plants that were under the aegis of those planets. Culpepper described: 1. The Way of making Plaisters, Ointments, Oils, Poultices, Syrups, Decoctions, Juleps, or Waters of all Sorts of Physical Herbs, that you may have them ready for your Use at all Times of the Year. 2. What Planet governeth every Herb or Tree used in Physic that groweth in England. 3. The time of gathering all Herbs, both vulgarly and astrologically. 4. The Way of drying and keeping the Herbs all the Year 5. The Way of keeping their Juice ready for Use at all Time. 6. The Way of making and keeping all Kinds of useful Compounds made of Herbs. 7. The Way of mixing Medicines according to the Cause and Mixture of the Disease and Part of the Body afflicted".

Another treatise "Adam in Eden" that is quoted by herbalists was written by William Coles in 1657. He opined that God through Adam the first Human according to the book of Genesis could name and find all creatures and plants according to their appearance, shapes, and other properties. Perusal of the above list shows that many of the herbs listed are also currently marketed in the various health food stores as herbal medicine for various ailments even though the conditions for which they are being prescribed does not necessarily match the descriptions of William Cole and without modern scientific analysis.

The Astronomer Giambattista Della Porta described with illustrations various plants with heart shaped leaves, fruits, or bulbs for heart disease, thistles, catkins, and scaly plants as treatment for skin diseases in his book Phytognomonica published in 1538. In the new world, physicians named Martin de La Cruz and another by name Badianus described various new world species of plants used by Aztec Indians which were considered to have medicinal value.

The Gardening tradition: John Gerard (1545-1612) who was an avid gardener wrote the Herbal or Generall Historie of Plantes wherein he described the morphology of many plants including some considered to have medicinal value. His book was revised later by Thomas Johnson in 1633 with descriptions of more than 2000 medicinal plants. Another herbalist John Parkinson wrote "Paradisi in sole Paradisis terrestris" and "Theatum Botanicum" in 1629 and 1640 respectively wherein, he listed many useful plants, but he also used the Doctrine of Signatures as a guide in some instances.

The Shaker tradition: In the USA, the Shakers popularized the use of medicinal herbs for treating various diseases by setting a thriving business in cataloguing and selling several plant products from native American and imported European plants for treating diseases. Some of the plants such as thorn apple, black henbane they marketed were toxic.

Comments: Although most of the plants identified by the systems described above had no true modern scientific basis, it served to classify plants and in some cases proved to be of value if not for what they were originally prescribed for but for other applications. However, because of the current interest in alternative medicine many of the above plants in one form or other continue to find a place in Health Food Stores as remedies for different ailments.

Zoopharmacognacy: Self-medication by animals: Anthropologists, Ethno botanists and others have been investigating the self-medicating behavior of animals particularly primates wherein they protect themselves from diseases caused by parasites and other microorganisms. Most animals also suffer from the very same types of diseases that humans are inflicted by whether they are caused by microorganisms or genetic conditions. Animals have been noticed to chew on or eat specific herbals or even dirt or charcoal. It is now being recognized that we can learn from animal behavior and use the information therefrom to find plants that may have some medicinal value. Thus, for example, a chimpanzee was seen to chew on the pith of a small tree known as the bitter leaf tree *Veronia amygdalina* although this *tree* is not a food tree for the monkeys. On further observation, it was noted that this Chimpanzee, which was sick earlier, became fully active. In addition, the dung of the Chimp was found to have the intestinal parasite *Oesophagostomum stephanostomum*. Another monkey, which chewed on the pith of this tree, was found to be free of this parasite. This observation led to the finding that this tree had the compound Veronioside B1 that had anti-parasitic, anti-bacterial, and anti-tumor properties. What was most interesting was the finding that the leaves of this tree had several other poisonous compounds, which were perhaps harmful to the monkey, but the pith that the monkey chose had only minimal quantities of the other toxic compounds while it was rich in the anti-parasitic Veronioside. Thus, the monkey had figured out which part of the plant was useful for its good health! As it turns out, the local people also use the pith and other parts of this plant in regulated doses to treat parasitic infections including the deadly Schistosomiasis caused by Trypanosomes. Now it is known that *Vernonia amygda*lina has seven steroid glucosides, as well as four sesquiterpene lactones, capable of killing parasites that cause schistosomiasis, malaria, and leishmaniasis. The sesquiterpene lactones (previously known to chemists as "bitter principles") are not only anthelmintic but also antiamoebic, antitumor, and antimicrobial.

Another documented example is the observation that Capuchin monkeys in Costa Rica chew on bark, leaves of *Piper marginatum*, *Clematic dioica*, and *Sloanea terniflorastems*, and rub the juice obtained on their bodies to repel insects. These plants do have insect repellent properties. Leaves of plants like *Rubia cordifolia, Lippia plicata, Ficus exasperata*, and *Aneilema aequinoctiale* were swallowed by Apes in Africa and the local populations eat these leaves to get rid of parasites, and stomach ailments. By inference, the local people had found these plants by seeing animal behavior. Holly Dublin of the Wild Life Fund reports that African elephants eat leaves of a tree from the Boraginacea family in order to induce labor and native women from Kenya brew a tea from the leaves of this plant to induce labor. Another anecdotal report from Karen Strier of the University of Wisconsin states that muriqui monkeys from Brazil eat the leaves of *Apuleia leiocarpa* and *Platypodium elegans* which contain compounds similar to estrogens which may help to decrease fertility and similarly they eat the fruits of *Enterlobium contortisiliquim* which contain progesterone like compounds which increases the chances of fertility. The Wild Ginger, which is found to have antibacterial properties, is consumed by Gorillas, and is sold in local African markets as a medicinal berry.

Several other drugs such as quinine, coca, Ipecac have been discovered by observation of various native populations. Thus, Jesuit priests saw native Indian tribes using the bark of the Cinchona tree for treating Malaria. Similarly, the discovery of Cocaine/Novocain as painkillers resulted from observation of the Indians chewing on leaves of the Coca plant for toothache. Similarly, the discovery of curare tubo-curarine chloride as an anesthetic, Ipecac as an emetic and anti-amoebic drug came about because of seeing native populations in South America consuming plant parts. Similarly, in Trinidad and Tobago, people use extracts from the following plants for treating snakebites: *Bauhinia cumanensis, Aristolochia rugosa, Pithecellobim unguis-cati, Aristolochia trilobata, Cola nitida, and Cecropia peltata*. Whether these plants are of any real use for treatment of snakebites must be evaluated. It is likely that the use of these plants by these people came about by seeing animal behavior in combination with trial and error. On the other hand, according to popular folk lore, the mongoose is supposed to protect itself from snakebites by eating the roots of *Rauwolfia serpentina* The Indian Snake root plant. Although, this is untrue, this belief and the common use of the roots of this plant to control mental ailments by local populations in India instigated further study of the pharmacological properties of this plant and helped to identify the beneficial effects of this plant as a source for lowering blood pressure and for its calming effects. The study of self-medication by animals could reveal the potential uses of many hitherto unknown for medicinal purposes but detailed follow-up studies are needed before any plants from such studies are found to be truly medicinal.

Other Examples

- Bears, deer, elk, and various carnivores, as well as great apes, are known to consume medicinal plants to self-medicate.

- Some lizards are believed to respond to a bite by a venomous snake by eating a certain root to counter the venom.

- Fruit flies lay eggs in plants having high ethanol levels when they detect parasitoid wasps, a way of protecting their offspring.

- Red and green macaws, along with many animals, eat clay to aid digestion and kill bacteria.

- Female woolly spider monkeys in Brazil add plants to their diet to increase or decrease their fertility.

- Pregnant lemurs in Madagascar nibble on tamarind and fig leaves and bark to aid in milk production, kill parasites, and increase the chances of a successful birth.

- Pregnant elephants in Kenya eat the leaves of some trees to induce delivery.

Comments: The doctrine of signatures, observation of animals and native populations such as aboriginal people, ethnic populations and tribal have served to advance progress in identifying plants as a source of medicines for improving human and animal health. However, caution is urged while using these herbal drugs since they may have other toxic products that may be harmful.

References

Web-Internet

https://www.britannica.com/topic/Ebers-papyrus/media/177583/127826

https://www.ancient.eu/The_Vedas

https://www.sciencedirect.com/topics/agricultural-and-biological-sciences/shennong

https://www.ncbi.nlm.nih.gov/pmc/articles/PMC4267359

http://exhibits.hsl.virginia.edu/herbs/badianus

http://exhibits.hsl.virginia.edu/herbs/herball

http://www.k https: //www.nlm.nih.gov/hmd

http://www.keralaayurvedics.com

https://www.britannica.com/biography/John-Gerard

https://www.sciencedirect.com/topics/agricultural-and-biological-sciences/vernonia-amygdalina

Books and Journals

Allegretti SM, et al. (2014): The use of Brazilian medicinal plants to combat Schizostoma mansoni Bagher Rokni M, editor Schistosomiasis 2012 www.intechopen.com/download/get/type/pdfs/id/25967

Arber, A. (1953), Herbals, Their Origin and Evolution, In Cambridge University Press, United Kingdom, pp 326 (https: //doi.org/10.1017/CBO9780511711497.010)

Bennett, Bradley C. (2007). "Doctrine of Signatures: An Explanation of Medicinal Plant Discovery or Dissemination of Knowledge?" Economic Botany, 61 (3): 246–255. doi: 10.1663/0013-0001(2007)61[246: DOSAEO] 2.0.CO, 2. ISSN 0013-0001 Retrieved 2008-08-31.

Benzie, I.F.F. and Wachtel-Galor, S., (Eds) (2011) Herbal Medicine, Clinical and Biomolecular aspects, CRC Press, Boca Raton, Florida

Boehme, J. (1621). Signatura rerun or the signature of all things Retrieved from http: //sacred-texts.com/eso/sat/index.htm

Buchanan, Scott (2014). The Doctrine of Signatures, A Defense of Theory in Medicine. United Kingdom: Taylor & Francis p. 142. ISBN 0415614155

Dalal, R., (2010), Hinduism, an alphabetical guide, Penguin books, de Roode JC, Lefèvre T, Hunter MD (2013). Ecology Self-medication in animals, Science 2013, 340(6129): 150–151.

Emmart, Emily Walcott, (1940), THE BADIANUS MANUSCRIPT; (Codex Barberini, Latin 241) Vatican Library, An Aztec Herbal of 1552, Published by Johns Hopkins, Baltimore, 1940, pp 341. (Introduction, Translation, and Annotations by Emily Walcott Emmart)

Fruth B, et al. (2014): New evidence for self-medication in bonobos: Manniophyton fulvum leaf- and stemstrip-swallowing from LuiKotale, Salonga National Park, Democratic Republic of Congo, Am J Primatology 2014, 76(2): 146–158.

Gerard's Herball (1927) edited by Marcus Woodward (a selected reprint of the 1636 edition).

Huffman M. (1997): Current evidence for self-medication in primates: A multidisciplinary perspective Yearbook Phys Anthronol 1997, 104(suppl 25): 171–200.

Journal of the History of Medicine and Allied Sciences (different volumes)

Lefèvre T, et al. (2012) Evidence for trans-generational medication in nature. Ecology Letters 2010, 13(12): 1485–1493.

Sreeman Namboothiri, D., (1994): Yogamrutham (Ayurvedam), Vidyarambham publishers, Mullakkal, Alleppey, Kerala, India. (In Malayalam), pp 412,

Walter Lack, H: (2001): Ein Garten Eden by and translated by Martin Walters.

Warrier, A.C., (1995): Ashtangahridayam (Ayurvedam), 6th edition, Devi Book Store, Kodungalloor, Kerala, India, pp 355 (Book in Malayalam language).

Worthm, J. E. (1991), The shakers and their proprietary medicines, Bulletin of the History of Medicine, Vol. 65, No. 2 (SUMMER 1991), pp. 162-184, Published by: The Johns Hopkins University Press

CHAPTER 3

Ethno-Pharmacy

All societies in the world whether in Africa, Asia, Europe, or America and even the subarctic regions have depended on plants as a source of not only food, fiber, and fuel but also for medicinal purposes. Many have developed their own individual systems of medicine based on plants. These include the Ayurveda, Siddha, Unani and tribal systems of India, the Traditional Chinese Medicine (TCM), Kampo (Japan), Tibetan (Sowa-Rigpa medicine), Sri Lankan ("Hela wedakama" and Desheeya Chikitsa) and Myanmar systems similar to Ayurveda and other indigenous forms, the African, Australian aboriginal and American Indian ethno-medicine, Egyptian, Greco-Roman and Arab-Islamic medicine and herbals of Tropical rainforest dwellers. Careful examination shows that these entire systems share certain plants that are common to all, but some are specific. Brief outlines of these ethno medical systems with focus on plant products are given below.

The purpose of this chapter is to show and find the role of plants in Ethno pharmacy and is not a comprehensive account of these important medical systems that are the basis of alternative medicine. The reader should refer to advanced texts for more information of these systems of medicine.

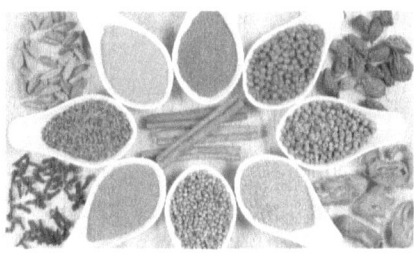

Ayurveda: Indians believe that the knowledge of medicinal plants is older than history itself, that it was gifted hundreds of thousands of years ago to the original inhabitants by Brahma, the divine creator himself. Thus, when the sages of the Ayurveda looked to heal human suffering, they were able to draw on knowledge that had already been evolving for millennia in the forests of India. This is a very ancient system practiced in India even today and dating back to 3000-5000 years ago. Mention and descriptions of this system have been made in all the four holy scriptures of India known as the Vedas. Initially, this system known as "Shruthi" was transferred orally from the teacher to the disciples and passed on from generation to generation. Subsequently, full details were written down in Sanskrit by Sushruta in the Sushrutha Samhita 3rd-4th Century CE, by Charaka in the Charaka Samhita 3-4th century CE, The Bheda Samhita Bower transcripts of 6th century, the Ashtanga Hridaya of Vaghabhatta 6th Century, Kashyapa and Hartha Samhita 6th Century of the Gupta period. The Bower manuscripts (or Yashomitra Manuscript, is now in the collections of the Bodleian Library in Oxford) mention the names of sage's Ātreya, Hārīta, Parāśara, Bhela, Garga, Śāmbavya, Suśruta, Vasiṣṭha, Karāla, and Kāpya as contributors to the understanding of medicinal plants listed there in. These texts were written on birch bark or specially

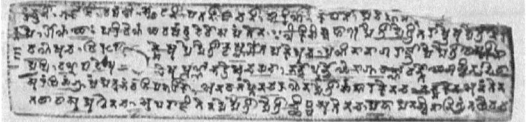

treated palm leaves that surprisingly survived and decoded by experts. Several other authors embellished on these texts and wrote detailed descriptions of this system and the plants used for treatment. These include Madhava Nidana by Madhavakara 7th Century, Ratirahasya by Kokkoa 9th Century, Sarangadhara Samhita by Sharangadhara 14th century and Bhavaprakasa by Bhava Mishra 15th century and Ananga Ranga by Kalyanamalla 16th century, Vaidya rathnavali by Raja Serfoji 1798-1832. Recent textbooks by Krishnan Kutty Varrier on History of Ayurveda and Ayur arogya sowkyam in Malayalam Ayurvedic Grandham and other texts in various regional languages give reviews and lists of medicinal herbs.

The term Ayurveda is derived from two Sanskrit words Ayur meaning life and Veda meaning Science. It is a composite system that states that Life is due to the synergistic effects of five elements namely Fire, water, air, earth and space (The Pancha bhutas) and that all individuals are composed of three constituents namely Vata, Pitta and Kapha collectively known as thri-doshas. However, based on individual characteristics individuals show dominance in one of these Doshas and will then show proneness to certain types of diseases. Thus, Vata governs all movement related diseases, pitta governs digestion, body temperature, nutrition, and metabolism and Kapha governs flesh/mucous. All three must be in balance. If out of balance, it will show specific symptoms that the Ayurvedic physician can diagnose and prescribe concoctions of plant based Ayurvedic formulations.

The Ayurvedic system divides diseases into eight types: 1. General medicine (Kaaya Chikitsa), 2. Pediatrics (Baala or Kaumara-bhrtya), 3. Surgery, (Shalya Chikitsa/tantra) 4. Treatment of Diseases of the eye, ear nose throat, neck and head, (Urdhvanga chikitsya), 5. Demonology/psychiatry, (Bhoota Vidya/Graha Chikitsya), 6. Toxicology, (Agatha Tantra/Dharmshta Chikitsya), 7. Virility/Aphrodisiacs (vajikarana tantra), 8. Geriatrics (Jara Chikitsya). The advanced nature of this system is indicated by the classification of tissues (Dhatus) into seven types and the specific treatments for ailments affecting these seven types of tissues that are plasma (rasa), blood (raktha), muscles (mamsa), fat, bone (asthi), marrow (majja) and semen (shukra).

This science based the diagnosis of diseases on eight different diagnostic techniques namely: pulse, urine analysis, analysis of stool, appearance of tongue, speech, touch, vision, and general appearance. Thus, one can see that Ayurveda understood human anatomy and used most of modern medical diagnostic techniques. Several recent molecular genetic studies show that the classification of individuals based on the thri-doshas (three characters) has a molecular basis. Thus, Single nucleotide polymorphism (SNP) studies do indeed verify classification of individuals into the three Doshas. These studies have attempted to correlate Prakriti classification with genetic information and association of single nucleotide polymorphisms (SNPs) in HLA-DRB17, CYP2C198, EGLN19, and inflammatory and oxidative stress related genes10, CD markers for various blood cells, DNA methylation alterations and risk factors of cardiovascular or inflammatory diseases. More studies that are recent have elaborated and verified the prakriti classification using SNP studies.

For treatment, Ayurveda used a holistic approach consisting of nutrition/diet, exercise and treatment using many plant products individually and in combination and in some cases included inorganic metals like gold, copper, silver, mercury and animal products like milk, yogurt and clarified butter Ghee and also charcoal and clay. Many of the herbals used in

Ayurveda were also the same as that used in Greco-Roman, Chinese, and other systems but also had several plants that were unique to the flora of India. Plants like *Ephedra* (Ephedrine), *Curcuma longa* (Turmeric), *Zingiber officianalis* (Ginger), *Raulwfia serpentina* (Snakeroot), *Piper nigrum* (Pepper), *Withania somniferum* (Ashwagandha), *Azadirachta indica* (Neem) and *Salix* sp (Willow) are some examples of Ayurvedic herbs that have proven medicinal value. Ayurvedic preparations are usually a combination of several plant products since the principle of Ayurveda states that any disease is expressed through different symptoms and that treatment should involve the amelioration of all these symptoms while at the same time concentrating on eliminating the primary cause of disease. For example, a fever caused by a microorganism could manifest through elevation of temperature, inflammation, gastrointestinal disturbance etc and so the Ayurvedic preparation is compounded to treat all these symptoms in a single preparation containing many medicinal plant parts plus in some cases also animal products like yogurt, milk and butter.

Ayurvedic Medical Management Strategy

Ayurvedic principles of treatment take different approaches depending on the medical needs of a patient. The following are some treatment protocols.

Abhyanga: This is massage protocol using a warm premedicated herbal oil. The herbs in the oil are based on the needs of the specific condition.

Bashpasweda: In this herbal steam treatment the patient is seated in a steam chamber in which steam emanates from a boiling herbal decoction.

Basti: This is an herbal enema treatment that acts as a colon cleanser. This treatment refreshes the colon and allows the colon to be repopulated with good probiotics and removes any toxins that may have accumulated in the colon.

Dhanyamladhara: Dhanyamla treatment involves covering the patient with a preparation consisting of Navara rice, horse gram, millet, citrus fruits, and dried ginger and covering the body with a warm cloth, This Ayurvedic procedure is prescribed to combat obesity, inflammation, muscular pain, neuropathy, hemiplegia, and rheumatic complaints. The duration of the treatment is 45 to 50 minutes depending upon the condition of the patient. The procedure is named from the Sanskrit words Dhanya=grains, Amla= Phyllanthus emblica/Nellikkai/Amla and dhara=covering.

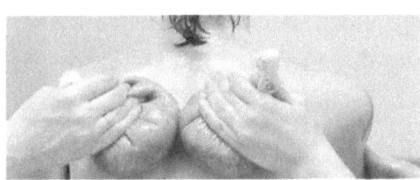

Dhara: This is a massage therapy involving rubbing the body with Herbal oil, coconut water, milk, ghee, Dhanyamla. The type of Dhara is an important therapy in Ayurveda and it gives sudden relief from chronic headaches, insomnia, mental tension, hysteria, hallucination, and insanity.

Garshana: The body is brushed with a dry wool or silk glove. This enhances circulation and cleanses the skin so that later oil and herbal treatments can penetrate deeply into freshly cleansed pores.

Nasya: This is one of the treatments involved in rejuvenating the sensory organs and in this case, it is the nasal region. Herbal oil is dripped into the nasal passages helping to clear he sinuses and drain the sinus. This process combats dryness of the nasal membranes and allows smooth breathing.

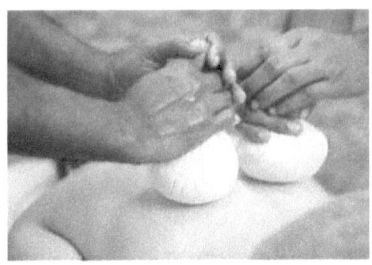

Navara Kizhi massage: Navara is a variety of medicinal rice. This rice is cooked in cow milk and the cooked rice is then wrapped into a bundle known as Kizhi. This kizhi is then dipped in warm medicated oil and the kizhi is then used as a massaging unit and the whole body or parts being treated is massaged with this kizhi. Usually people who have become paralyzed after stroke or injury or even infections are treated thus to bring back the affected part of the body back to functionality. Recovery is a slow process and several days of such massage would be prescribed to get the desired results.

Pinda Swedana: This is a treatment to rejuvenate the skin and is part of the sensory treatment protocol. Rice cooked in milk and herbs is massaged into the skin tissues and joints. The treatment is deeply relaxing and rejuvenating.

Shiro-Abhyanga-Nasya: This is a combinatorial treatment using medicated oil and steam inhalation to clear the head and neck region as well as nasal passages. The process consists of deep head/neck/shoulder massage and facial lymphatic massage, followed by deep inhalation of therapeutic aromatic steam, and a nasal and sinus Nasya with herbalized nose drops.

Panchakarma: Panchakarma is a natural therapy that detoxifies and restores the body's inner balance and vitality. There are five different treatments involved here and depending on the diagnosis, the Ayurvedic physician prescribes only some of the five treatments or all five. The five treatments are: 1. Abhyanga, 2. Shirodhara, 3. Garshana, 4. Swedana, and 5. Udvartana.

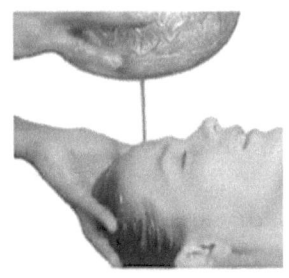

Shirodhara: This Ayurvedic therapy involves gently pouring or dripping medicated liquids usually oil from a hanging pot with a tiny hole over the forehead and can be one of the steps involved in Panchakarma.

Snehapana: This is a pre-purification therapy before Panchakarma treatment. Herein, the patient is made to drink clarified butter known as ghee prepared from cow milk. The patient is also given an oil massage with medicated oil.

Swedana: Here, the patient is placed in an herbal steam bath such that the head and the heart regions are kept cool and the body is heated by steam with herbal medicinals.

Udvartana: Herbal paste or herbal is rubbed on the lymph nodes under the neck, arms, and legs to stimulate the lymph tissues to restore best discharge.

Vamana (induced vomiting): Specific herbal extracts are given to patients to expel increased phlegm from the body.

Virechana: This is part of the Panchakarma treatment for detoxification that involves using plant medicines that have a laxative effect, mainly aimed at reducing pitta dosha (Bile acids) and toxic accumulation in the gastrointestinal tract, liver, and gallbladder.

Yogavasti: A herbal medication is given by enema, aiding in diminishing extra vata dosha present in the body.

A list of some prominent Ayurvedic herbs is listed in **Table 2 below**. This list shows the botanical and popular Ayurvedic names of the plants along with their phytochemistry, parts used and therapy.

Table 2: List of selected Ayurvedic herbs with their Botanical and Sanskrit names and usage with comments

Botanical Names	Ayurvedic and popular name	Family	Phytochemicals	Parts Used	Therapy &notes
Acorus calamus	Haimavati, bhutanashini, jatila,/Sweet Flag	Acoraceae	alpha-asarone or beta-isoasarone	Rhizome	Sedative, diuretic, hallucinogen, The FDA prohibits calamus use in food products and is labeled Likely Unsafe
Adhatoda vasika./ Justicia adathoda	Vaasaka/Adalodam	Acanthaceae	Vasicine	Leaf	good for voice, cough, fever
Allium cepa	Palaandu/Onion	Alliaceae	Quercetin	Bulb	rheumatism, strength promoting, aphrodisiac
Allium sativum	Rasona/Garlic	Alliaceae	Allin	Bulb	aphrodisiac, bone fracture, eye health, rejuvenation, cough, heart health
Aloe vera	Ghrita Kumari / Aloe	Xanthorrhoeaceae	Mannans, lectins, polymannans, anthraquinone, C-glycosides, anthrones,	Leaf	Skin protection, burn treatment, ulcer prevention and healing in the digestive system Some studies show carcinogenicity in rats.
Anisomeles malabarica	Sprikkaa	Labiatae; Lamiaceae	Beta-sitosterol	Whole plant	Aphrodisiac, skin diseases, pruritus
Andrographis paniculata	Kalmegh, bhumi-neem/Nila vembu/	Acanthaceae	Andrographolide, Paniculide	Leaves, root, rhizome	Colitis, joint pains including joint pains due to Dengue fever
Apium graveolens	Ajamodaa/ Celery	Apiaceae; Umbelliferae	d-Lemonene	Seed	Digestive stimulant, wholesome for heart, semen promoting, Causes kidney stones.

Aquilaria agaliocha	Krumija, jagdhaAnaryaka	Thymeliaceae	Agarospirol, Aquillochin	Wood oil	Ayurvedic oil used in various neuro-muscular conditions
Areca catechu	Guwaaka/ Betel nut	Arecaceae; Palmae	Alkaloids, tannins	Fruits	Digestive stimulant
Artemisia vulgaris	Damanaka	Asteraceae/ Compositae	Aristolochic acid	Whole plant	wholesome for the heart, aphrodisiac, fevers
Asparagus racemosus	Shatavari/ Asparagus	Asparagaceae	Saponins	Roots	Rejuvenate, brain tonic, aphrodisiac, strength promoting, complexion enhancing, Digestive-stimulant, galactagogue
Azadirachta indica	Nimba/Neem	Meliaceae	Rutin	Leaf	Skin diseases, polyuria, and worm infestation
Bacopa monnieri	Brahmi/Indian pennywort	Plantaginaceae	Baccosides	Whole plant	-memory enhancer, skin diseases, anemia, and edema
Baliospermum montanum,	Danti	Euphorbiaceae	Terpenes	Root	piles, i calculus, colic, pruritus, abdominal diseases
Bauhinia purpurea	Kovidaara	Caesalpiniaceae	Quercetin	Flower, seed,	worm infestation, skin diseases), prolapse of rectum,
Bauhinia variegata	Kaanchanaara	Caesalpiniaceae	Tannins	Flower	worm infestation, skin diseases, prolapse of rectum,
Benincasa hispida Syn B. cerifera	Kuushmaanda	Cucurbitacea	Beta-sitosterol	Fruit	Disorders of urinary bladder, peptic ulcers
Boerhavia diffusa	Punarnavaa	Nyctaginaceae	Punarnavoside	Whole plant	Improving Kidney functions, liver diseases

Continued...

Species	Common name	Family	Active constituent	Part used	Uses
Cannabis sativa	Bhangaa Marijuana	Cannabinaceae	Delta-9-tetra-hydrocannabinol	Leaf	Digestive stimulant, narcotic, pain killer, multiple conditions
Carthamus tinctorius	Kusumbha/Safflower	Asteraceae	Saponin	Aerial part	Dysuria, hemorrhagic disorders
Carum carvi	Krishna jiraka	Umbelliferae	Carvone	Fruit	Digestive stimulant, aphrodisiac,
Cassia fistula	Aaragvadha	Caesalpiniaceae	Fistulin	Flowers and pods	purgative, fever, diseases of heart,
Cassia occidentalis Linn	marda	Caesalpiniaceae	Anthraquinone glycosides,	All	aphrodisiac, anorexia,
Cassia tora	Chakramarda	Caesalpiniaceae	Anthraquinones	Seed and leaf	Ring worm, skin diseases, pruritus.
Centella asiatica	Manduukaparni/Gotu kola	Umbelliferae;	Asiaticoside, asiatic acid	Whole plant	Increase life span, brain tonic, Skin diseases, anemia, and edema.
Chenopodium album	Vaastuuka	Chenopodiaceae	Ascaridole	Leaf and seed	digestive stimulant, semen promoting, worm infestation
Cinnamomum camphora	Karpura/Camphor	Lauraceae	Camphor volatile oil	Leaf	Chest congestion, joint pain
Cinnamomum tamala	Patra	Lauraceae	Cinnamaldehyde, linalool	Bark Leaf	Digestive stimulant,
Cinnamomum zeylanicum/C. Verum	Daarusitaa/Cinnamon	Lauraceae	Cinnzeylanin	Leaf, bark, and root	semen promoting, strength promoting.
Clitoria ternatea	Aparaajitaa/Blue pea	Fabaceae	Kaempferol	Seed, bark, root, leaf	intellect promoting, beneficial for throat, vision, and memory
Coleus Forskohlii/C. barbatus	Mayani/Indian coleus	Lamiaceae	diterpenoid -forskolin	Root	Glaucoma, weight loss, stimulates insulin, Cardio benefits. recent studies show severe side effects

Commiphora mukul	Guggul/Indian bdellium-tree	Burseraceae	Guggulsterone Z and E	Resin from Bark	Lowers cholesterol and triglycerides
Convolvulus pluricaulis	Shankhapushpi	Convolvulaceae	Sankhpushpine alkaloids	Whole plant	Brain tonic, aphrodisiac, strength promoting, epilepsy, worm infestation, skin diseases
Coriandrum sativum	Dhaanyaka/celantero	Umbelliferae;	Delta-linalool	Fruit	digestive stimulant,
Crocus sativus	Kumkuma/Saffron	Iridaceae	Crocin	Dried style and	Complexion promoting, diseases of head
Cuminum Cyminum	Jiraka/Cumin	Umbelliferae;	Cuminaldehyde	Fruit	Digestive stimulant, anti-flatulent; aphrodisiac,
Curcuma angustifolia	Tikhuri	Zingiberaceae	d-ar-Curcumene	Dried rhizome	Disorders of blood, urticaria, hyperacidity, anorexia, dyspepsia, Thirst
Curcuma longa	Haridraa	Zingiberaceae	Turmerone	Dried rhizome	Skin health, polyuria, edema, antioxidant, immunostimulant
Curcuma zedoaria	Karchuura	Zingiberaceae	Curcumenol	Dried rhizome	digestive stimulant, cough, piles, ulcer
Desmostachya bipinnata, Syn: Eragrostis cynosuroides	Dharbha, Kusha grass/Halfa grass	Gramineae (Poaceae)	Not evaluated	Leaves, stem	Asthma, kidney stone, diarrhoea, wound healing, histamine properties, known to hyper-accumulate radioactive caesium-137 a byproduct of nuclear fission.

Continued...

Eclipta alba	Bhringaraaja/false daisy	Asteraceae; Compositae	Wedelactone, Demethyl wedelolactone	Whole plant	beneficial for hair and skin, diseases of eye,
Elettaria cardamomum	Sukshmailaa/cardomom	Zingiberaceae	1,8-Cineol	Seed fruit	Cough, piles, dysuria.
Embelia ribes	Vidanga/false black pepper	Myrsinaceae	Embelin	Fruit	digestive stimulant, worm infestation, colic, tympanitis
Emblica officinalis	Amalaki/Indian gooseberry	Euphorbiaceae	Phyllembin/gallic acid	Fruit leaf seed	Geriatrics, aphrodisiac, polyuria
Ephedra gerardiana	Soma	Ephedraceae	Ephedrine	Aerial part	Rejuvenating and geriatric
Ferula foetida,; Syn F. assafoetida	Hing/asafoetida	Apiaceae; Umbelliferae	Asaresinotannols, ferulic, ferulic, ferulic acid	Oleogumresin	Digestive stimulant, colic, diseases of abdomen, constipation, worm infestation
Ficus religiosa	Ashvattha/Sacred Fig	Moraceae	Beta-sitosterol	Bark fruit	Diseases of female genitalia, burning sensation,
Flacourtia ramontchi,; Syn F. indica	Vikankata	Flacourtiaceae	Flacourtin	Bark and leaf	Snake bite, Arthritis, Throat infections
Foeniculum vulgare	Mishreyaa/Fennel	Apiaceae; Umbelliferae	Anethole, fenchone	Fruit	Digestive stimulant, a digestive, colic
Fumaria parviflora/F. indica	Parpata	Fumariaceae	Protopine	Whole plant	fevers, blood disorders, chronic skin diseases, urinary diseases, and cough
Garcinia indica; Syn G. purpurea Roxb	Vrkshaamla/Mangosteen	Guttiferae; Clusiaceae	Garcinol, isogarcinol, hydroxycitric acid	Fruit, root, bark, Leaf	piles, sprue, abdominal colic, diseases of heart/weight loss

Glycyrrhiza glabra	Yashtimadhu/ Licorice	Fabaceae; Papilionaceae	Glycyrrhizin, glycyrrhizin acid	Root	Eye health, aphrodisiac, beneficial for hair, voice, ulcers, edema, poison, vomiting, phthisis
Gymnema sylvestre	Meshashringi/ Sugar destroyer	Asclepiadaceae	Gymnemagenin	Leaf	Cough, polyuria, pain in the eyes, Digestive, stimulant, purgation, Diabetic control
Hedychium spicatum; Syn *H. album*	Shati/ginger lily	Zingiberaceae	Hedychenone, 7 hydroxyhedychenone	Rhizome	Cough,
Hemidesmus indicus, Syn *Periploca indica* Linn.	Saarivaa/Nannari/ Indian Sarsaparilla	Asclepiadaceae;	Hemidesmine, hemidesmin-1	Root	Semen promoting, dyspepsia, cough, excessive vaginal discharge, acute diarrhea, prostate health
Hibiscus abelmoschus, Syn *A. moschatus*	Lataakasturi/ Okra/Ladies finger	Malvaceae	Beta-sitosterol	Seed	aphrodisiac, beneficial for eyes, expectorant, diseases of the urinary, bladder, stool softener.
Hibiscus rosa-sinensis Linn.	Japaa/Shoe Flower	Malvaceae	Cyanidin-3-sophoroside	Flower	beneficial for hair/sunscreen
Hyoscyamus niger	Khuraashaanikaa/ black henbane	Solanaceae	Hyoscine, hyoscyamine	Whole plant	digestive stimulant, Madakaari intoxicant. Causes hallucinations Tachycardia
Momordica charantia	Kaaravellaka/ Karela/Bitter gourd	Cucurbitaceae	Momordicosides, charatin	Fruit and seed	anemia, polyuria, worm infestation, Diabetes, bowel movements
Moringa oleifera, Syn *M. pterygosperma* Gaertn.	Shobhaanjana/ Drumstick	Moringaceae	Spirochin, niazirin	Whole plant	colic, skin diseases, Kshaya phthisis, Swasa.

Continued...

Mucuna monosperma	Kaakaandolaa	Fabaceae; Papilionaceae	Mucunine	Seed	Source of L-DOPA
Mucuna pruriens, M. prurita	Kapikachhuu/ velvet bean	Fabaceae; Papilionaceae	Mucunine	Seed and root	aphrodisiac, strength promoting. Cause itching
Musa paradisiaca, M. sapientum Linn.	Kadali/Table Banana	Musaceae	Sitoindoside IV	Flower	aphrodisiac, bulk promoting,
Myristica fragrance	Jaatiphala/True nutmeg	Myristicaceae	Myristicin	Endosperm	digestive stimulant, beneficial for the voice,
Nardostachys jatamansi/N. grandiflora DC.	Jataamaansi/Kaanti Vardhak/muskroot	Valerianaceae	Jatamansone	Root	brain tonic, promotes luster, strength promoting, skin diseases.
Nerium indicum; Syn N. odorum Soland.	Karavira	Apocynaceae	Digitoxigenin	Root and leaf	skin diseases, inflammation, piles, worm infestation, and abortifacient.
Nyctanthes arbor-tristis	Paarijaata/Night Jasmine	Oleaceae; Nyctanthaceae	Irridoid glycosides	Leaf	edema, obesity, worm infestation, ear diseases
Nymphaea alba.	Kumuda/Lotus	Nymphaeaceae	Nymphalin	Flower and seed	complexion enhancing, blisters, erysipelas.
Nymphaea stellata	Nilotpala/Blue Lotus	Nymphaeaceae	Quercetin	Leaf	retention of urine, diseases of the nervous system, insanity, epilepsy,
Ocimum basilicum, O. caryophyllantum, O. minimum, O. pilosum	Barbari/Basil	Labiatae; Lamiaceae	Estragole	Whole plant	wholesome for heart, digestive stimulant,

Ocimum sanctum Linn. Syn *O. tenuiflorum*.	Thulasi/ Holy Basil	Labiatae; Lamiaceae	Carvacrol	Whole plant	cardiac diseases, digestive stimulant, dysuria, and intercostal neuralgia,
Papaver somniferum	Ahiphena/Opium poppy	Papaveraceae	Morphine, isoquinoline alkaloids	Seed and poppy	cough, intoxicant, pain killer, sedative
Phyllanthus niruri	Bhuumyaamalaki/ stone breaker	Euphorbiaceae	Niuride	Whole plant	cough, bleeding disorders, pruritus.
Picrorhiza kurroa.	Katuki	riaceae	Picroside kutcoside	Root	stool softening, polyuria, cough, skin diseases, helminthes.
Piper betle	Taambula/betel leaf	Piperaceae	Chavicol	Leaf, fruit	strength promoter, digestive stimulant.
Piper chaba; P. retrofractum, P. officinarum	Chavya	Piperaceae	Piperine	Fruit and root	digestive stimulant, a digestive, diseases of the anal canal.
Piper cubeba	Kankola/tailed pepper	Piperaceae	Cubenine	Fruit	wholesome for heart, showed in loss of vision
Piper longum	Pippali/ Long Pepper	Piperaceae	Piperine	Fruit	Rasayana rejuvenative, cough, Shvaash, brain tonic, digestive stimulant.
Piper nigrum	Maricha/Black pepper	Piperaceae	Piperine	Fruit	digestive stimulant, relieving colic, anthelmintic.
Plectranthus barbatus syn *Coleus forskohlii*	Maayinmula/Indian Coleus	Lamiaceae (Asterds)	Foksolin, rosmarinic acid		heart disease, convulsions, spasmodic pain, and painful urination. Most popular use now is as a controversial miracle weight loss pill.

Continued...

Botanical name	Common name	Family	Chemical constituents	Part used	Uses
Pongamia pinnata; Syn P. glabra, P. indica	Karanja/Pongam oil tree	Papilionaceae; Fabaceae	Karanjin, quercetin	Seed	Diseases of female reproductive system, skin diseases, worm infestation, insecticide, piles.
Punica granatum	Daadima/Pomegranate	Punicaceae	Granatin A and B	Fruit, root, bark.	brain tonic, strength promoting, fever, diuretic, stomach ailments
Putranjiva roxburghii; Syn Drypetes roxburghii	Putranjiva	Euphorbiaceae	Putranjivosides, glucosinolates	Fruits, seed, kernel	aphrodisiac induces conception.
Rauvolfia serpentina	Sarpagandhaa/Indian snakeroot	Apocynaceae	Reserpine	Root	Snake bite, scorpion sting, spider bite, rat bite, fever, worm infestation, ulcers, mental illness, hypertension
Ricinus communis	Eranda/Castor bean	Euphorbiaceae	Ricinoliec acid	Seed	colic, edema, pain in bladder, diseases of abdomen, fever, hydrocele, cough, skin diseases, rheumatism. Seeds have ricin cell poison
Saraca asoca,; Syn S. indica	Ashoka/Ashoka tree	Caesalpiniaceae	Quercetin	Bark	complexion promoting, leucorrhea, thirst, burning sensation, worm infestation, emaciation, poison control
Sesamum indicum Linn; Syn S. orientale Linn.	Tila/Sesame	Pedaliaceae	Sesamin, sesamolin	Seed	Beneficial for hair, beneficial for skin.
Swertia chirayita, S. chirata,	Chiraayita/felwort	Gentianaceae	Chiratin glycoside, decussatin	Whole plant	Febrifuge,
Syzygium aromaticum	Lavanga/Cloves	Myrtaceae	Eugenin, eugenol	Flower buds	digestive stimulant, colic, cough, hiccup, mouth freshener

Syzygium cuminii; *Eugenia jambolana*	Jambu/Java plum	Myrtaceae	Bergenin	Bark and seed	Diabetic control
Tamarindus indica; *T. occidentalis T. officinalis*	Amlika/Tamarind	Caesalpiniaceae	Tamarindienal	Fruit	Digestive stimulant.
Taxus baccata	Sthauneyaka Taxus/Yew	Taxaceae	Taxol	Leaf and bark	obesity, itching, skin diseases, rejuvenative, foul smell of the body, cancer
Tephrosia purpurea, T. hamiltonii	Sharapunkhaa/Wild indigo	Fabaceae; Papilionaceae	Rutin	Whole plant	Anti-helminthic, fish poison
Terminalia arjuna.	Arjuna/Arjun tree/ Marudha maram	Combretaceae	Arjunatosides	Bark	wholesome for heart, obesity, polyuria, ulcer.
Terminalia bellirica	Vibhitaki/beleric/ Thaanikkai	Combretaceae	Bellaricanin, beta-sitosterol	Fruit	Beneficial for hair, cough.
Terminalia chebula Retz.	Haritaki/Chebula myrobalan/ Kadukkai	Combretaceae	Terflavin, Many glycosides	Fruit	rejuvenating, cough, laxative
Tinospora cordifolia	Guduuchi/ Chittamruthu	Menispermaceae	Berberine, columbine	Stem	Anti-diabetic
Trachyspermum ammi, T. copticum, Carum copticum,	Yavaani/Ajwan/ caraway	Apiaceae; Umbelliferae	Thymol	Fruit	antiseptic, antispasmodic, aromatic, diaphoretic, digestive, diuretic, expectorant, tonic, anti-gas
Tribulus terrestris	Gokshura/caltrop	Zygophyllaceae	Tribuloside	Fruit	Digestive-stimulant, aphrodisiac, increases testosterone

Continued...

Botanical name	Common name	Family	Chemical constituents	Part used	Medicinal uses
Trigonella foenum-graecum	Methikaa/Fenugreek	Fabaceae; Papilionaceae	Trigonelline	Seed	Pacifies and useful in fever, anti-diabetic.
Valeriana wallichii; V. jatamansi, Nardostachys jatamansi	Tagara/Indian valerian	Valerianaceae	Beta-bargamotene	Rhizome	Epilepsy, diseases of eye, sedative.
Vetiveria zizanioides,	Ushira/Vettiver	Poaceae;	Khusilal	Root	Digestive, dysuria. It grows in oil-contaminated soil (soil remediation?)
Vitex negundo Linn.	Nirgundi/Chinese chaste tree	Verbenaceae	Casticin, beta-sitosterol	Leaf and root	Hair, beneficial for eyes, Worm infestation, rheumatism.
Vitis vinifera Linn.	Draakshaa/Grapes	Vitaceae	Anthocyanin	Fruit	Beneficial for eyes, bulk promoting, digestive stimulant.
Withania somnifera	Ashwagandhaa/Indian Ginseng	Solanaceae	Withanine, somniferine	Root	geriatric, intellect-promoting, strength-promoting, semen promoting, leucoderma, edema.
Zingiber officinale	Aardraka/Ginger	Zingiberaceae	Zingiberene, gingerol	Rhizome	Digestive stimulant, carminative, cough, edema, elephantiasis, piles.

Siddha System: It is said that the Siddha system was in prevalence even during the days of the Indus valley civilization and that the Ayurvedic system used many of the same principles as the Siddha system. This system is said to have originated from the sage Agasthya who had in turn received instructions from the God Muruga who is highly revered in the state of Tamil Nadu in India. Agasthya in turn passed on the secrets of this mode of medical system to 18 different people who came to be known as Siddhas or those who know the secrets of perfect health or ultimate bliss. The 18 Siddha's are Thiruvarutchitthar, Thirumoolar-siddhar, Kolangi-siddhar, Bogar-Siddhar, Pulippani-Siddhar, Sattamuni-Siddhar, Macchamuni-Siddhar, Ramadheva-Siddhar, Kamalamuni-Siddhar, Sivavaakiya-Siddhar, Karuvoor-Siddhar, Edaikaadar-Siddhar, Sundharanadhar-Siddhar, Azhukanni-Siddhar, Paambatti=Siddhar, Therayar-Siddhar, Konganar-Siddhar and Korakkar-Siddhar.

All these Siddhars or Siddhas were Saivite Tamils who worshipped Lord Siva and have authored books on various aspects of Siddha system of medicine The Siddha system was then passed on from the teacher Guru to the student and until recently were the domain of individual families. However, during the last several years, due to the efforts of the State Government of Tamil Nadu, India, palm leaf inscriptions which have survived the test of time containing the formulations of the Siddha system have been deciphered albeit partially and schools and colleges have been established to educate and train students in this system.

Like Ayurveda, Siddha system is also a holistic system based on the principle that all health conditions result from the status of Vatha during childhood, during adulthood and Pitha during old age. As in Ayurveda and Greco-Roman medicine, Siddha medicine also is based on the idea that all activities in the human body result from Solid-Earth, Fluids-water, radiance/heat-Fire, gas-air and ether space-divinity/soul. The drugs used by Siddhars for treatment include a combination of Herbs (Thavaram), inorganic substances (Thadhu) and animal products (jangamam). Many Siddha medicines have Arsenic, Gold and Mercury prepared in very secretive procedures. Thus, alchemic ideas pervaded Siddha medicine. Although, mercury and arsenic are highly toxic elements, Siddha practitioners argue that they treat these elements in such a way to make them innocuous except as medicines. Modern Chemistry does not support this. The common preparations of Siddha medicines are Bhasma (calcined metals and minerals), churna (powders), Kashaya (decoctions), Lehya (confections) Ghrita (ghee/clarified butter preparations) and Taila (oil) preparations. Siddha have specialized in Churna (metallic preparations) which become alkaline, mezhugu (waxy preparations) and Kattu preparation that are impervious to water and flames. Diagnosis is again based on pulse, urine, touch, appearance, eyes, skin color, and appearance of tongue and examination of faeces that are like allopathic techniques in internal medicine.

The plant parts used include roots, bark, leaves and flowers individually or in combination. The medicinally important plants share commonality with Ayurveda but have greater focus on the flora of southern India. The following lists some of the plants that have medicinal value as per Siddha medicine. Most of these are plants that grow in the Southern parts of India. As in the case of Ayurveda, Unani and TCM, the Siddha system makes use of combinations of different herbs and sometimes alchemy to prepare herbal formulations for different health conditions.

Table 3: List of plants of medicinal value in the Siddha system: The medical claims are not necessarily backed by controlled experimental or clinical data

Botanical name	Siddha/ Common Name	Family	Parts	Uses/Comments
Aegle marmelos	Koovilam, Vilvam or Bael, Wood Apple	Rutaceae	Fruit	Dyspepsia, sinusitis, appetite modulator
Allium sativum	Poondu, or Garlic	Amaryllidaceae	Bulb	Diabetes, hypertension
Alternanthera Sessilis	Ponnanganni, Dwarf Copper leaf	Amaranthaceae	Leaves	Hyper- acidity, skin conditions
Andrograhis paniculata	Nilavembu, Kālamegha	Acanthaceae	Rhzome. root, leaves	upper respiratory infection, ulcerative colitis, rheumatism, joint pain (Dengue joint pain)

Botanical Name	Common Name	Family	Part Used	Uses
Amaranthus Polygonoides	Sirukeerai	Amaranthaceae	leaves	voice improvement, mouth and stomach ulcers, dysuria
Amaranth	Mulai keerai	Amaranthaceae	leaves	dysuria
Asparagus racemosus	Taneervittan kizhangu, Shatavari	Asperagaceae	Phylloclades	Gastric ulcers, galactagogue.
Annona squamosa	Krishnagaru/ Sugar apple	Annonaceae	Fruit, seeds	Headache, Migraine
Borreria hispida	Nathaichuri	Rubiaceae	Leaves	Emetic, Febrifuge, diarrhoea, dysentery, hemorrhoids
Carica papaya	Pappali, Papaya	Caricaceae	Fruit	Digestive aid, Abortifacient
Cardiospermum halicacabum	Mundakkathan, Balloon plant	Sapindaceae	Fruits	Diahorrea
Cassia auriculata	Avarai or Meghari	Fabaceae	Leaves, Flowers	Diabetes, Skin
Cassia Senna	Ponnavarai, Senna	Fabaceae	Leaves	Laxative
Catharanthus roseus	Nithya kalyani, Madagaskar Periwinkle	Apocyanaceae	Flowers, leaves	cancer, (Vincristine, Vinblastine) diarrhoea and diabetes
Celosia argentea)	Pannai keerai	Amaranthaceae	Leaves	ulcerative colitis
Centella asiatica	Vallarai/Brahmi	Apiaceae	Leaves	Memory, skin, ulcerations, eczema,
Chrysopogon zizanioides	Vettiver	Poaceae	Roots	Skin care, aroma therapy
Cinnamomum Zeylanicum	Lowangapattai		Bark	Stimulant, carminative, aphrodisiac.
Cissus quardrangularis	Pirandai	Vitaceae	Stem	tonic, analgesic, heal broken bones

Continued...

Scientific Name	Common Name	Family	Part Used	Uses
Citrullus colocynthis	Kummattikkai	Cucurbitaceae	Fruit	Diabetes, cause intestinal distress
Cocos nucifera	Thengai	Palmaceae	Coco-Water	Electrolyte, prevents dehydration
Cocos nucifera	Thengai	Palmaceae	Milk	Mouth ulcers, Skin
Cocos nucifera	Thengai	Palmaceae	Oil	Effective moisturizer. Prevents flaking and dryness of skin,
Cocos nucifera	Thengai	Palmaceae	Flower paste in milk	Removes Kidney stones
coleus aromaticus	Karpooravalli	Lamiaceae	leaves	Expectorant, controls cough
Curcuma longa	Manjal/Turmeric	Zingiberaceae		Diabetes, cancer, Immune booster
Cympopogon jwarankusa	Vilamicham ver	Cyperaceae	roots, rhizome	Fever, joint pain of Dengue fever
Cuminum siminum	Jeera or Cumin:	Apiaceae	Seeds	Flatulence, Hyper acidity
Cyanodon dactylon	Arugam pul/ Bermuda grass	Cyperaceae	Leaves, stem	Detoxification, Eye drops, blood clotting in wounds, diuretic
Cyperus rotandus	Korai kizhangu/ Kizha Nelli	Cyperaceae	Rhizome	Anti-microbial, anti-malarial, antioxidant, anti-diabetic,
Eletteria cardomum	Elakkai/ Cardomom	Zingiberaceae	Seeds	stomach problems, indigestion
Euginea jambolena/ Syzygium cumini	Naaval/Jambol	Myrtaceae	Fruit	Digestive ailments, Diabetic control
Euphorbia hirta	Ammanpacharisi	Euphorbiaceae	Leaves	Cough, chest congestion, impotency

Foeniculum vulgare	Perunjeerakam/ Fennel	Umbelliferae	Seeds	Digestive aid
Gymnema sylvestre	Sarkarrai kolli	Apocyanaceae	Fruit, leaves	Diabetes
Hibiscus rosa sinensis	chembaruthi	Malvaceae	Flower	Heart and hair health, diabetes
Indigofera tinctoria	Naranokki	Fabaceae	Leaves	Treats gray hair
Ionidium Suffruticosum	oridazh thamarai	Violaceae	Roots, Leaves	Sexual diseases, Libido
Mangifera indica	Mangai	Anacardiaceae	Flowers	mouth sores, stomach ulcers
Mentha arvensis	Pudina	Lamiaceae	Leaves	Hyper acidity
Mimosa pudica	Thottarchurungi (touch-me-not)	Fabaceae	Roots, Leaves	anti-cobra venom, anti-cancer, kills filarial worm larvae
Momordica charantia	Pakal kai, Pavakkai	Cucurbitaceae	Fruit, flowers	Diabetes, vermifuge
Moringa Oleifera	Murunga/ Drumstick	Moringaceae	Leaves, Fruit, Flowers	Cardiotonic, Anti-cancer, Diabetics
Myristica fragrans	aathikkai or Jaathikkai	Myristeceae	Fruit	Carminative, Digestive, Tonic, Aromatic
Ocimum sanctum	Thulasi/Holy basil	Lamiaceae	Leaves	Ailments of chest including Asthma, cough, cold, bronchitis, skin
Oldenlandia corymbosa	Parpatakam	Rubiaceae	Leaves	Febrifuge, Diaphoretic, Stomachic, Laxative, Joint pains, Dengue pain
Opuntia ficus-indica, Opuntia sp.	Kalli/Prickly pear	Cactaceae	Flower, stem	Hemorrhoids, Anti-fungal

Continued...

pentatropis capensis	Uppilankodi	Apocyanaceae	Leaves	analgesic and anti-inflammatory
Phyllanthis emblica	Nellikkai	Phyllanthaceae	Fruit	Gastritis, digestive problems. Hyperacidity, Hair tonic
Phyllanthus niruri	Kizha Nelli	Phyllanthaceae	Leaves	Disorders of the Liver and urinary tract, kidney stones,
Rauwolfia serpentina	Chivan melpodi	Apocyanaceae	Whole plant	Sedative, Anti-high blood pressure
Rumex acetosella	Pulicha keerai/ Gonkura	Polygonaceae	Leaves	digestive disorders, skin diseases, jaundice, tonic
Solanum nigrum	Manathakkali keerai	Solanaceae	Leaves, fruit	acidity control, stomach/mouth ulcers
Solanum Trilobatum	Thuthuvalai	Solanaceae	Leaves, fruits, roots,	cold, cough, Asthma
Solanum Xanthocarpum	kandankathiri	Solanaceae	Fruit	Cough, cold, breathing problems
Syzygium aromaticum	Clove or Elavangam	Myrtaceae		Hyperacidity, tooth ache
Terminalia arjuna	Neermaruthu	Combretaceae	Bark	Astringent, Coolant, Cardiac Tonic
Terminalia Belerica	Thanrikkai	Combretaceae	Nuts	Cardiovascular conditions
Terminalia chebula	Kadukkai	Combretaceae	Fruit	Laxative, piles, skin diseases
Tinospora cardifolia	Seenthil	Menispermaceae	Leaves	Osteoporosis, immune-modulation, anti-type 2-diabetes

Trachyspermum ammi	Ajwain/Carom	Apiaceae	seeds	Flatulence
Tribulus terrestris	Cherunerinche; Nerinji mul	Zygophyllaceae	Leaves, Nutlets	Hepatic diseases, diuretic, demulcent, aphrodisiac, muscle building, raising testosterone
Trichopus zeylanicus	Arogyapacha	Dioscoreaceae	Leaves, Flowers, Fruit	Tonic, Immune &Vision enhancer, Aphrodisiac, Anti-obesity.
Trichosanthes dioica	Peypudal, Kambupudal	Cucurbitaceae	Fruit	Febrifuge, anti-inflammatory
Foeniculum vulgare	Perunjeerakam/ Fennel	Umbelliferae	Seeds	Digestive aid
Trigonella foenum grecum	Mendyam/Methi	Fabaceae	Seeds	Diabetes
Tylophora indica	Nanjaruppaan	Apocyanaceae	Leaves, roots	bronchial asthma
Withania sominifera	Amukkura	Solanaceae	Leaves, berries, roots	Sedative, ulcers
Ziziphus jujuba	Seemai Ilanthai	Rhamnaceae	All parts	inflammation, hypotension,

Unani System: The Unani system is a Greco-Arabic-Persian-Islamic system of medicine which arrived in India in the 13th century after the establishment of the beginnings of Islamic rule in India and which became indigenized by including Ayurveda and Siddha systems. The teachings of Hippocrates, Galen, al-Razhi, al zhahrawi and Ibn Sena Avicenna and the Ayurvedic teachings of Sushruta and Charaka influenced the formulation of Unani medicine. Here also, the treatment options are based on the belief that all humans are composed of fluids-water, air, fire, and solids-earth but not ether. As in Ayurveda and Siddha medicine, the interaction of or imbalance of Phlegm, bile ~pitta and blood ~Vata form the basis of disease. The medical practitioners are known as Hakims. These Hakims used the canonical belief that imbalance of Phlegm, blood, and bile resulted in disease. In common with Chinese medicine, Unani system includes many animal products such as blood, bone powder, powder of the horn of animals besides minerals and herbs.

The following is a list of medicinal herbs used in Unani medicine. These botanicals are used in combination with minerals and sometimes-animal products.

Arjuna/yellow myrobalan (*Terminalia sp*), Amla (*Embelica officianalis*), Ajwan (*Ptychotys ajowan*), Senna (*Cassia augustifolia*), Sesame (*Sesamum indicum*), Black Pepper (*Piper nigrum*), Coriander (*Coriandrum sativum*), Licorice (*Glycyrrhiza glabra*), Violet (*Viola odorata*), Rose-Gulab (*Rosa sp.*), Ceylonese Cinnamon (*Cinnamomum zeylanicum*), Ginger (*Zingiber officianalis*), Adathodum (*Adathoda Vasica*), Lotus (*Nymphaea lotus*), Rhubarb (*Rheum emodi*), Cumin (*Cuminum siminum*), Cardomum (*Eletteria cardomum*), Neem (*Azadirachta indica*), Aatrilal (*Ammi majus*), Anjir (*Ficus carica*), Atis (*Aconitum heterophyllum*), Azaraqi (*Strychnos nux-vomica*), Badam (*Prunus amygdalus*), Baqla (*Vicia faba*), Bazrulbanj (*Hyoscyamus niger*), Chashmizaj (*Cassia absus*), Chobchini (*Smilax china*), Fifil Siyah (*Piper nigrum*), Gaozaban (*Borago officinalis*), Gulnar Farsi (*Punica granatum*), Habbun Neel (*Ipomoea nil*), Halela Siyah(*Terminalia chebula*), Heel Kalan (*Amomum subulatum*), Heel Khurd (*Elettaria cardamomum*), Heeng (*Ferula foetida*), Hina (*Lawsonia inermis*), Inderjao Shireen (*Wrightia tinctoria*) Inderjeo Talkh (*Holarrhena antidysenterica*), Ispand (*Peganum harmala*), Kaiphal (*Myrica esculenta*) Karanj (*Pongamia pinnata*), Karanjwa (*Caesalpinia crista*), Kasni (*Cichorium intybus*), Khar-e-Khasak (*Tribulus terrestris*), Khatmi (*Althaea Officinalis*), Khella (*Ammi visnaga*), Khulanjan (*Alpinia galanga*), Khurfa (*Portulaca oleracea*), Khurfa Khurd (*Portulaca quadrifida*), Kishneez Khushk (*Coriandrum sativum*), Konch (*Mucuna cochinchinensis*), Kulthi (*Dolichos biflorus*) Madar (*Solanum nigrum*), Mawiz (*Vitis vinifera*), Methi (*Trigonella foenum-graecum*), Mundi (*Sphaeranthus indicus*), Nakhud (*Cicer arietinum*), Narjeel Daryaee

(*Lodoicea maldivica*), Nilofar (*Nymphaea alba*), Panwad (*Cassia tora*), Palas (*Butea monosperma*), Sambhalu (*Vitex negundo*), Saranjan (*Colchicum luteum*), and Turbud (*Operculina turpethum*).

Tibetan (Sowa-Rigpa medicine) and other South Asian systems: Tibetan medicine is also an orally transmitted system that considers diet, behavior, medicines, and physical therapy as the four main methods of treatment of diseases. However, recently information about this system is being documented in writing by the Tibetan monks who migrated to India along with the Dalai Lama by the Yuthog foundation and by western researches. Here also importance is given to the role of the three humors and their interplay as in Greek, Ayurvedic and Chinese traditional medicine (TMC). The doctor goes through a visual and tactile examination of the patient and checks on the status of the patient's essential nutrient status. Importance is given to diet, behavior patterns and physical exercise. Besides, Buddhist tantrism and Ayurvedic treatment techniques based on use of plant-based products play a big role in treatment. The Ladakh region of India, Tibet, Bhutan, and Nepal follow the same types of holistic treatment options. Most of the herbal medicinals are formulated from plants grown in the Himalayan region.

Sri Lanka, Myanmar, Malaysia, Thailand, Cambodia, Indonesia and other South east Asian countries: Most of these countries had a connection with India either through conquests by Hindu kings or spread of Buddhism and hence these indigenous systems have much in common with Ayurvedic and Siddha formulations based on plants grown in these regions. Similarly, these countries were also influenced by Chinese Traditional medicine. The Khmers of Cambodia combined local herbal and spiritual beliefs with Indian, Vietnamese, and Chinese methods. In Vietnam, the Traditional Vietnamese medicine (Y học Cổ truyền Việt Nam), also known as Southern Herbology (Thuốc Nam) is popular and influenced by TCM.

Sri Lankan traditional medicine is a mix of Sinhala desheeya chikitsa practices/Sinhala Vedakama, or "Hela Vedakama" derived from Ayurveda and Siddha systems of India, Unani from Arabs, and Greek medicine. This system is also influenced by Buddhist and Tibetan systems of traditional medicine. Indonesian local medicine was also influenced by Arabic-Islamic protocols due to trade with Arabic merchants and later conversion of most of the population to Islam. All herbals used indigenously are the same as in Ayurveda/Siddha Unani and Arabic-Islamic systems along with other native folk medicinals based on plants that grow in abundance or unique to these countries.

In Myanmar (Old Burma) traditional medicine is divided into 1. The Desana system based on natural occurrence such as hot and cold. Its concepts are based on Buddhist Philosophy, with the therapeutics use of herbal and mineral compounds and diet. 2. The Bethitzza system, based on Ayurvedic concepts with extensive use of herbal and mineral compounds, 3. The Astrological system, based on the calculations of zodiac of stars, planets and the time of birth and age. These calculations are linked to prescribed dietary practices. 4. The Vezzadara system, dependent on meditation and practices of alchemy.

Traditional Chinese Medicine (TCM): Herbal products have been listed and used in Chinese traditional medicine for over 5000 years. Emperor Shennong Ben Cao Jing 2700 BCE described about 365 plants, which have medicinal uses including Ephedra -ephedrine, Licorice, Rhubarb, opium and Ginger in his treatise "Shen-nung-Pen-tsao. Huangdi known as the Yellow emperor wrote the classic of internal medicine known as "Huandi neijing" which stays even today as the foundation of Chinese traditional medicine. Zhang Zhongjing who lived in the first century of the Christian era has described 112 medically relevant herbs based on the theory of yin and Yang. The Chinese based their identification of herbals on empirical observations. In the Yin/Yang system the Yang which is the male principle is bright and heavenly while the Yin is the female principle which is dark and earthly. Diseases of the body are apportioned on the balance of Yin and Yang in the body. As in most other ancient systems of medicine the human body is said to be composed of five aspects namely wood, fire, earth, metal, and water and is influenced by five planets. Chinese ancients recognized only five planets, five colors and five conditions of the atmosphere. This is different from the Indo sub-continental system, which considers Air, Fire, Water, Earth, and space as the Pancha bhutas or five elements. However, Chinese medicine also considered the effect of the environment on the human body and gave it a spiritual significance. The Chinese also based their diagnostics of human health based on a cosmic system wherein it is postulated that there are 12 meridian channels in the body known as jingmai and 15 collateral channels known as luomai in addition to sub channels through which life energy known as qui flows and which play an important role in human health. Knowledge of these points is the basis of the science of acupuncture. This belief is akin to the Indian belief that there are seven chakras that are the abode of specific energy centers that branch into sub channels that control activities of various organs. However, there is no anatomical evidence for the presence of such channels. The Chinese system also recognizes five organs namely heart, lungs, liver, spleen, and kidneys in one category and another five that are in the viscera namely stomach, intestines, gallbladder, and urinary bladder. This separation of organs is like those described in the Indian systems. Thus, either independently or more likely, information must have been communicated from one society to another.

The Chinese Materia Medica that has since been changed with new additions is a pharmaceutical text written by Li Shizhen in 1518–1593 AD during the Ming Dynasty of China. This has plant and animal products along with minerals, gems, and earth, as is the case in many other ancient systems. The Chinese recognize at least 1000 plants for which medicinal and nutritive values have been assigned. The following is a list of plants, which are considered to have medicinal value. The use of these plants singly or in combination with minerals, animal parts in the form of decoctions, tablets, powders, oils, ointments, syrups or jams for various treatment procedures were based on the yin/yang theory wherein these medical concoctions were supposed to raise or lower yin or yang and therefore restore the body to good health. The following is a brief list of herbs with their use in Traditional Chinese Medicine (TCM).

Birthwort (*Aristolochia*): Used for treating hypertension, hemorrhoids, and colic. Recent studies show that it has safety issues because it causes cancer of upper urinary tract and kidney failure as well as Balkan endemic nephropathy.

Camellia (*Camellia sinensis*): is used in TCM for aches and pains, digestion, depression, detoxification, energizer, and prolongation of life.

Chinese cucumber (*Trichosanthes kirilowii*): treat tumors; reduce fevers, swelling and coughing, abscesses, amenorrhea, jaundice, and polyuria. It is extremely toxic if improperly used.

Chrysanthemum flowers (Ju Hua; (*Chrysanthemum sp*): used in TCM to treat headaches, fever, dizziness, and dry eyes.

Cocklebur fruit (*Xanthium*, cang er zi): Used in TCM, to treat sinus congestion, chronic nasal obstructions and discharges, and respiratory allergies. Xanthium is now known to be toxic and causes vomiting, diarrhea, and abdominal pain.

Crow dipper (sheng ban Xia, *Pinellia ternata*): Is used in TCM for removing phlegm even though crow dipper is highly toxic to humans. The constituents of this herb include methionine, glycine, ß-amino butyric acid, gamma-amino butyric acid, alkaloids 1-ephedrine and trigonelline, phytosterols and glucoronic acid.

Croton seed (*Croton tiglium*): is used in TCM to treat gastrointestinal disorders, convulsions, and skin lesions. It is often used with rhubarb, dried ginger and apricot seed, Croton has cancer-causing chemicals.

Dioscorea Root (*Radix Dioscorea*, Huai Shan Yao or Shan Yao in Chinese): is used to clear congestion of the lungs and fortify health of spleen and kidney.

Ginger root (*Zingiber officinale*): has been used in China for over 2,000 years to treat indigestion, upset stomach, diarrhea, and nausea. Ginger is also an important herbal in Ayurveda, Unani, and Siddha systems of Indian medicine.

Ginkgo (*Ginkgo biloba*): seeds are crushed to treat asthma. Ginkgo has been used in TCM for 5,000 years. Ginkgo is also considered good for memory retention. Further studies are needed to substantiate all these medical claims.

Ginseng root (*Panax sp.*): This is the most widely sold traditional Chinese medicine. The name "ginseng" is used to refer to both American (*Panax quinquefolius*) and Asian or Korean ginseng (*Panax ginseng*). Siberian ginseng or Eleuthero (*Eleutherococcus senticosus*) is different from Chinese/Asian Ginseng. Multiple health benefits have been attributed to Ginseng.

Goji berry/Wolf berry (*Lycium barbarum*): Potential benefits against cardiovascular and inflammatory diseases, vision-related diseases (such as age-related macular degeneration and glaucoma). Wolfberry leaves may be used to make tea, together with Lycium root bark (called dìgǔpí; in Chinese), for traditional Chinese medicine (TCM).

Horny goat weed (*Epimedium spp., Yin Yang Huo,*) is believed to be an aphrodisiac.

Lily bulbs (Bai He, *Lilium sp***)** are used to treat dry cough, sore throat, and wheezing.

Round Cardamom Fruit (Bai Dou Kou, *Eletteria cardomom*) is used to treat poor appetite,

Thunder Vine (lei gong teng, *Radix tripterygii wilfordii*) is used to treat arthritis, relieve pain, and reduce joint swelling. It can be extremely toxic,

Trichosanthis Root (*Radix Trichosanthis* or Tian Hua Fen in Chinese): is used, in the wasting and thirsting syndrome.

Strychnine tree seeds (*Strychnos nux-vomica*, Ma Quan Zi) are marketed and sold with a claim to treat diseases of the respiratory tract, anemia, and geriatric complaints. It has toxic strychnine, so can also be used as a poison for killing rodents.

Sweet wormwood (*Artemisia annua*, Qing Hao) is believed to treat fever, headache, dizziness and prevent bleeding. The plant has Artemesinin that has anti-malarial properties. The 2015 Nobel Prize was awarded for the discovery of the anti-malarial properties of Artimisin from this herb used in TCM as a treatment for parasitic diseases.

Willow bark (*Salix sp*): Willow bark has been used throughout the centuries in China and Europe to the present for the treatment of pain (particularly low back pain and osteoarthritis), headache, and inflammatory conditions such as bursitis and tendonitis.

Table 4: List of 50 plants used in TCM: Traditional Chinese medicine (TCM) also lists the following 50 plants in addition to the above as among the most important herbs used in TCM.

Botanical name	Chinese Name	Popular name	Botanical Family
Agastache rugosa	huò xiāng	Korean Mint	Lamiaceae
Alangium chinense	bā jiǎo fēng	Chinese Alangium	Cornaceae
Anemone chinensis	bái tóu weng	NA	Ranunculaceae
Anisodus tanguticus	Shan làng dàng	Whitleya Sweet	Solanaceae
Ardisia japonica	zi Jin niú	Primulaceae	Marlberry
Aster tataricus	Zi wǎn	Asteraceae	Tatar Aster
Astragalus propinquus	huáng qí	Fabaceae	Chinese Astragalus
Camellia sinensis	chá shù	Theaceae	Tea
Cannabis sativa	dà má	Cannabaceae	Marijuana/Hemp
Carthamus tinctorius	hóng huā	Asteraceae	Safflower
Cinnamomum cassia	ròu gùi	Lauraceae	Cinnamon
Cissampelos pareira	xí shēng téng	Menispermaceae	Velvet leaf
Coptis chinensis	duǎn è huáng lián	Ranunculaceae	Chinese Goldthread
Corydalis yanhusuo	yán hú suǒ	Papaveraceae	Chinese Poppy
Croton tiglium	bā dòu)	Euphorbiaceae	Purging Croton
Daphne genkwa	yuán huā	Thymelaeaceae	Lilac Daphne
Datura metel	yáng jīn huā	Solanaceae	Devil's Trumpet
Datura stramonium	zǐ huā màn tuó luó	Solanaceae	Jimson Weed
Dendrobium nobile	shí hú	Orchidaceae	Noble Dendrobium
Dichroa febrifuga	cháng shān	Hydrangeaceae	Chinese Quinine

Ephedra sinica	cǎo má huáng	Ephedraceae	Chinese Ephedra
Eucommia ulmoides	dù zhòng	Eucommiaceae	Hardy rubber tree
Euphorbia pekinensis	dà jǐ	Euphorbiaceae	Peking spurge
Flueggea suffruticosa	yī yè qiū	Phyllanthaceae	
Forsythia suspensa	liánqiáo	Oleaceae	Weeping Forsythia
Gentiana loureiroi	dì dīng	Gentianaceae	
Gleditsia sinensis	zào jiá	Fabaceae	Chinese Honey locust
Glycyrrhiza uralensis	gān cǎo	Fabaceae	Licorice
Hydnocarpus anthelminticus	dà fēng zǐ	Achariaceae	Chaulmoogra tree
Ilex purpurea	dōngqīng	Aquifoliaceae	Purple Holly
Leonurus japonicus	yì mǔ cǎo	Lamiaceae	Chinese motherwort
Ligusticum wallichii	chuān xiōng	Apiaceae	Szechwan lovage
Lobelia chinensis	bàn biān lián	Companulaceae	Creeping Lobelia
Phellodendron amurense	huáng bǎi	Rutaceae	Amur cork tree
Platycladus orientalis	cè bǎi	Cupressaceae	Chinese Arborvitae
Pseudolarix amabilis	jīn qián sōng	Pinaceae	Golden Larch
Psilopeganum sinense	shān má huáng	Rutaceae	Naked rue
Pueraria lobata	gé gēn	Fabaceae	Kudzu
Rauwolfia serpentina	shégēnmù	Apocyanaceae	Indian Snakeroot
Rehmannia glutinosa	dìhuáng	Orobanchacea	Chinese Foxglove
Rheum officinale	yào yòng dà huáng	Polygonaceae	Chinese or Eastern rhubarb
Rhododendron tsinghaiense	Qīng hǎi dù juān	Ericaceae	Rhododendron
Saussurea costus	yún mù xiāng	Asteraceae	Costus root
Schisandra chinensis	wǔ wèi zi	Schisandraceae	Chinese Magnolia Vine
Scutellaria baicalensis	huáng qín	Lamiaceae	Baikal Skullcap

Continued...

Stemona tuberosa	bǎi bù	Stemonaceae	
Stephania tetrandra	fáng jǐ	Menispermaceae	Stephania Root
Styphnolobium japonicum	huái	Fabaceae	Pagoda Tree
Trichosanthes kirilowii	guā lóu	Cusurbitaceae	Chinese Cucumber
Wikstroemia indica	liāo gē wáng	Thymeliaceae	Indian stringbush

Korean Traditional Medicine: Korean Traditional medicine is a mix of Japanese and Chinese practices and incorporates several animal products as in TCM along with plant products. Korean medicine is a traditional East Asian medicine, it uses acupuncture, herbal medicine, and various other modalities as the main tools to promote health and treat disease. Korean medicine shares 80 percent of the theory, points, and techniques with Chinese medicine and Japanese medicine. Traditional Korean Medicine considers the human body as a microcosm of the universe. Here also as in Ayurveda and TCM, the basis of all ailments is due to the imbalance of five elements and treatment is holistic in nature. Many different herbs are used as dry powders, tablets, decoctions, and extracts. One of the most important herbs/trees used in KM is ginseng.

Japanese native Medicine: Known as **Kampo**, this system is also like TCM and is based on the principle that human health is based on the balance between physical body traits and mental makeup. In Japan, the administration of crude herbal drug formulations dates back by more than 1500 years. Recent decades have seen a revival of Kampo medicine in medical practice, accompanied by a scientific reevaluation and critical examination of its relevance in modern health care. The term "Kampo", which literally means "method from the Han period (206 BC to 220 AD) of ancient China", refers to its origin from ancient China. The basic therapeutic handbook for the application of herbal prescriptions was the Shang Han lun. During the Edo-period from 1600 onwards, the specific Japanese characteristics of Kampo took shape. The system reduced the number of medicines used in TCM to just 300. The treatment regime is based on symptoms. For the determination of the appropriate herbal prescription, the physician carries out a thorough investigation of the complaints and symptoms of the patient, including taking their temperature, examining sensation, weakness or sweating, symptoms which are not often primarily taken into account in conventional medicine. The physical examination includes abdominal palpation, tongue inspection and pulse diagnosis. The most important herbs found in majority of Kampo formulations are Licorice and Ginger.

African Traditional Folk Medicine: The continent of Africa is a storehouse of many yet unexplored or un-described plants. Being a big continent inhabited by many ethnic groups, each such group practices specific treatment options, which are only becoming clear now. Of more than 6400 species of plants, more than 4000 are used for medicinal purposes. The roots, stem, bark, leaves, and flowers are used. African medicine also regards human health is the result of interaction between body, spirit and mind and they also divide diseases as those due to physical causes and those due to spiritual possession by evil spirits.

The herbal preparations commonly used in Africa are also used by traditional healers in most countries where the African populations migrated voluntarily or during the slave trade.

Table 5: Medicinally important plants of Africa: The list is based on traditional uses and is not an endorsement for specific therapy. However, further modern experimentation may prove veracity of claims made by traditional herbalists.

Botanical name/Family	African traditional medicine for
Acalypha integrifolia: Euphorbiaceae	Intestinal worms
Achyranthesaspera: Amaranthaceae	diabetes, cancer, Diuretic, hepatoprotective, inflammation, Arthritis, antiparasitic, antioxidant, anti-hyperlipidemic, anti-mosquito larvicidal and lymphatic Filariasis, anti-microbial, Spermicidal, asthma, epilepsy, anti-depressant, and anti-obesity.
Acokanthera oblongifolia: Apocyanaceae	Treatments for snakebites, itches, and internal worms.
Aframomum corrorima: Zingiberaceae	Seeds are used as a tonic, carminative, and laxative.

Continued...

Asystasia gangetica: Acanthaceae	Used in Nigerian folk medicine for the management of asthma.
Aframomum melegueta: Zingiberaceae	In West African folk medicine, are valued for their warming and digestive properties.
Alafia barteri: Apocyanaceae	Local medicinal uses include treatment for malaria and rheumatism
Alafia lucida: Apocyanaceae	For jaundice, eye problems and stomach complaints
Alafia multiflora: Apocyanaceae	Treatment for wounds, ulcers, and abdominal pains
Alafia scandens: Apocyanaceae	Local medicinal uses include treatment for rheumatism
Alchornea floribunda: Euphorbiaceae	A decoction of the dried leaves is used to treat diarrhea and the leaves are pulped to promote the healing of wounds.
Aloe buettneri: Asperagales Asphodelaceae	The leaf extracts can be applied externally for skin conditions such as burns, wounds, insect bites, Guinea worm sores and vitiligo. In Côte d'Ivoire Togo, the roots are used as antimalarial. In Nigeria, leaf sap is used as anti-helminthic. Throughout West Africa, juice is used for liver problems
Asystasia gangetica: Acanthaceae	In Nigerian folk medicine for the management of asthma.
Baissea axillaris: Apocyanaceae. Apocyanaceae;	Treatment for kidney problems and colic and as a diuretic
Baissea multiflora: Apocyanaceae	The plant's many local medicinal uses include treatment for colic, rheumatism, arthritis, kidney problems, hemorrhoids, lumbago, conjunctivitis, appendicitis, diarrhea, and gonorrhea
Balanites aegyptiaca: Zygophyllaceae	Fruit is mixed into porridge and eaten by nursing mothers and the oil is consumed for headache and improved lactation. A decoction of the bark is also used as an abortifacient and an antidote for arrow-poison in West African traditional medicine. Bark extracts kill copepod hosts of Schizostoma.
Bauhinia petersiana: Fabaceae	In Zimbabwe, the Shona ethnic group uses the plants roots for treating dysmenorrhea and female infertility. In South Africa, the leaves are mixed with salt and pounded and are used to cure wounds. Also, in majority of countries the macerated roots of the plant are used for treating diarrhea
Begonia sutherlandii: Cucurbitaceae	Infusion of the leaf and stem of the plant is used medicinally by Zulu people in South Africa to treat heartburn and vomiting of blood

Bidens pilosa (umhlabangubo). Asteraceae	Treat infertility. In TCM, it is used to treat wounds.
Boscia senegalensis: Capparidaceae	Fruits seeds have about 25% protein and considered as a source of food under famine conditions
Boswellia dalzielii: Burseraceae;	Gallic and protocatechuic acids were isolated as the main antibacterial and antioxidant principles of the stem bark.
Bridelia micrantha: Phyllanthaceae	Anti-abortifacient, laxative, diverse conditions of the central nervous system (headache), eye (infections, conjunctivitis), the gastrointestinal system (abdominal pain, constipation, gastritis), respiratory system (common cold), and the skin (scabies) and used hygienically as a mouthwash.
Cannabis sativa (intsango), Cannabinaceae	Treat bronchitis, asthma, hypertension, labor pains, pain in general
Carduus tenuiflorus (uMhlakavuthwa). Asteraceae	The plant is used to extricate poison or diseases from a sick person. It is believed the plant sucks out the cause of the illness in itself
Catharanthus lanceus: Apocyanaceae;	Toothache, fever, skin diseases, and as a diuretic
Catharanthus trichophyllus: Apocyanaceae'	Sexually transmitted diseases, impotency, back pain, toothache, fever, dysentery, bleeding, and liver diseases. It is also used as a stimulant, aphrodisiac, and appetite suppressant
Cayratia debilis: Vitaceae	Fruits are inedible, although are fed to poultry in the Central African Republic to protect from influenza and coccidiosis.
Citropsis articulate: Rutaceae	In Uganda, an infusion made of the ground root, drunk once a day for three days is a powerful aphrodisiac for men only
Cochlospermum angolense: Bixaceae	The bark showed activity against the rodent malaria parasite Plasmodium berghei in laboratory tests. Traditional medicine known as borututu is sold in Africa. It may be a prospective herbal for human malaria also.
Craspidospermum: Apocyanaceae	Treatment for pulmonary diseases and syphilis
Cussonia spicata: Araliaceae	A bark decoction is used in the treatment of malaria, and indigestion in a similar manner to tonic water, A root decoction is used to treat fever, venereal disease, as well as a diuretic and laxative. A root bark decoction is used to treat mental illness
Cyclopia intermedia: Fabiaceae (honeybush)	The leaves are commonly used to make herbal teas. It grows only in small areas in the southwest and southeast of South Africa and has many similarities with rooibos tea (Aspalathus linearis).

Continued...

Dacryodes edulis: Burseraceae	Treat various ailments such as wound, skin diseases, dysentery, and fever. A wide range of chemical constituents such as terpenes, flavonoids, tannins, alkaloids and saponins have been isolated from the plant.
Deinbollia oblongifolia: Sapindaceae	The roots are used in traditional Zulu medicine for stomach complaints
Detarium senegalense: Caesalpineaceae/ Fabiaceae	Medicine for the removal of the placenta after birth and treatment of anemia, wounds, skin problems, bronchitis, pneumonia, stomachache digestive disorders, tuberculosis, and in cases of heavy blood loss. Root decoctions are used to treat marasmus, debility, intestinal complaints, and convulsions. Leaf and shoot, mixtures have been used in the treatment of dysentery, conjunctivitis, arthritis, fractures, and boils. Seeds have been effective in controlling blood-glucose levels in diabetic individuals, for the treatment of mosquito bites and as an antidote against arrow poison and snake bite
Dichrostachys cinerea: Fabaceae;	The bark is used for headache, toothache, dysentery, elephantiasis, root infusions are used for leprosy, syphilis, coughs, as an anthelmintic, purgative, and strong diuretic. Leaves are used for epilepsy and as a diuretic and laxative, and a powdered form is massaged on limbs with bone fractures.
Diplorhynchus: Apocyaneace	Treatment for indigestion, diarrhea, fever, snakebite, infertility, venereal disease, diabetes, pneumonia, and tuberculosis
Emex australis (inkunzane).	root is used on infants and adults suffering from restlessness or constipation
Entada rheedii, Fabaceae	Used to induce dreams. The plant is also used as a topical ointment against jaundice, toothache, ulcers and to treat muscular-skeletal problems
Erianthemum dregei: Loranthaceae	Treat stomach complaints in children and cattle
Erythrophleum africanum: Fabaceae	In Africa, the stem of the plant is used as a toothpick for oral hygiene because of the richness of the plant in flavonoids and anthocyanidins.
Euphorbia candelabrum: Euphorbiaceae	In traditional Ethiopian medicine, its sap was used as a purgative to cure syphilis, and when mixed with other medicinal plants as a salve to treat the symptoms of leprosy
Galenia secunda, Aizoaceae	Root extracts used to treat kidney pains
Garcinia kola: Clusiaceae (Rosids, Malphiales)	Purgative, antiparasitic, and antimicrobial. The seeds are used for treating bronchitis, throat infections, colic, head or chest colds, and cough. It is also used for liver disorders and as a chewing stick.

Helichrysum petiolare: Asteraceae;	The leaves and twigs are boiled and prepared as a sort of tea to soothe coughs and fever. The leaves are also applied to wounds to prevent infection and are ceremonially burnt to produce traditional incense.
Hoodia gordonii: Asclepediaceae;	Appetite Suppressant and weight loss. Popular in Herbal medicine shops
Hunteria congolanaL: Apocyanaceae	Fever, diarrhoea and as an anthelmintic
Hunteria umbellate: Apocyanaceae	fever, leprosy sores, stomach, and liver problems and as an anthelmintic, especially against internal worms
Hunteria zeylanica Apocyanaceae	Stomachache
Khaya anthotheca Meliaceae	The bark has a bitter taste, which is often used as a medicine for common colds. Green fruits are said to be poisonous Oil from seeds is rubbed into scalp to rid of insects and lice.
Kigelia Africana: Begoniacease	Dried or fresh fruits are used to deal with ulcers, sores, and syphilis - the fruit has antibacterial activity.
Leonotis leonurus, Lamiaceae	Known for its medicinal and mild psychoactive properties
Lonchocarpus laxiflorus Fabaceae;	A decoction of leafy twigs with those of *Byrsocarpus coccineus* is drunk and used as a lotion to treat dermatitis. The powder of calcined roots is rubbed on the forehead to treat headache.
Margaritaria discoideaese Phyllanthaceae	A leaf-decoction is taken in Ivory Coast and Chad for conjunctivitis and for poisoning. The bark is used as a purgative in West Africa and as an anthelmintic in Central Africa. The Fula people in Africa use the bark for toothache; in the Central African Republic, a decoction is used for post-partum pains, stomach and kidney complaints and ease parturition. In Malawi, the powdered bark extract is applied to swellings and inflammation for quick relief.
Microloma saggittatum Apocyanaceae	An infusion of the fleshy root fascicle of has been used in traditional medicine for relief of griping pains in the abdomen. The infusion is red in color.
Mikania natalensis: Asteraceae	Traditional Zulu and Swazi medicine for urinary complaints, headaches, backache, and colds.
Milicia excelsa: Moraceae	The powdered bark is used for coughs, heart problems and lassitude. The latex is used as an anti-tumor agent and to clear stomach and throat obstructions. The leaves and the ashes also have medicinal uses.

Continued...

Momordica balsamina: Cucurbitaceae;	Treat wounds
Momordica charantia (Bitter melon): Cucurbitaceae)	Used in various Asian and African herbal medicine system for stomach complaints. to relieve diabetes, laxative, anti-bilious, emetic, anthelmintic agent, treatment of cough, respiratory diseases, skin diseases, wounds, ulcer, gout, and rheumatism. It is contraindicated in pregnant women because it can induce bleeding, contractions, and miscarriage
Monodora myristica: Myristicaceae	Like nut meg. Used as stimulants, stomachic, for headaches, sores and as insect repellent.
Motandra guineensis: Apocyanaceae	Local medicinal uses include as a treatment for eye infections, toothache, headache, and postpartum stomach pain
Myrtus nivellei: Myrtaceae;	Rhinosinitis
Ochrosia oppositifolia. apocyanacease	Local medicinal uses include as a carminative and in high doses as an abortifacient. Leaves, bark, and the oil from the seed are used medicinally against baldness, ringworm, nosebleeds, chest complaints, eye infections, and venereal disease. Infusions of the bark are also used by Kenyan Maasai warriors to gain courage as well as an aphrodisiac and a blood-strengthening tonic. The root is used orally or as an enema and as a purgative for cattle.
Petchia madagascariensis: Apocyanaceae	Local medicinal uses include as a treatment for stomach-ache, gonorrhea, rheumatism, gout, malaria and as a diuretic and anthelmintic
Plectranthus amboinicuse: Lamiasceae;	Leaves for treatment of coughs, sore throats, and nasal congestion. In India used to treat malarial fever, renal and vesical calculi, cough, chronic asthma, hiccough, bronchitis, helminthiasis, colic, convulsions, skin ulcerations, scorpion bite, skin allergy, wounds, diarrhoea, with emphasis on the leaves being used as a hepatoprotective, to promote liver health. In Indonesia, it is a traditional food used in soup to stimulate lactation for the month or so following childbirth.
Pleiocarpa mutica: Apocyanaceae;	Local medicinal uses include as a treatment for stomach-ache, kidney diseases, malaria, jaundice and as a laxative
Pleioceras barteri: Apocyanaceae;	Local traditional medicinal uses include as an emmenagogue, abortifacient and in the treatment of rheumatism and malaria
Punica granatum: Lythraceae	Pomegranate ellagitannins: historically, used in African traditional medicine for antibiotic and antifungal purposes

Prunus Africana: Rosaceae	An herbal remedy prepared from the bark of P. africana, is used as an alternative medicine for benign prostatic hyperplasia (BPH). A Cochrane Review of the research concluded that a standardized preparation of P. africanum may be useful for lower urinary symptoms consistent with BPH
Pterocarpus soyauxii: Fabaceae.	Bark extracts are used in herbal medicine to treat skin parasites and fungal infections
Pycnanthus angolensis: Myristicaceae;	Eye washes to treat cataracts and Filariasis of the eye. The bark has been used as a poison antidote and a treatment for leprosy, anemia, infertility, gonorrhea, and malaria. Leaf extracts are consumed or used in an enema to treat edema. Root extracts are used to treat parasitic infections, such as Schistosomiasis. The seed oil is used to treat thrush.
Pycnobotrya: Apocyanaceae;	a treatment for chest infections, hematuria, diarrhoea, dysentery and bronchitis
Salix mucronata: Salicaceae:	Pain
Salvia africana-lutea: Lamiaceae	Was used by early European settlers to treat colds, tuberculosis, and chronic bronchitis. Traditional indigenous healers use it for respiratory ailments, influenza, gynecological complaints, fever, headaches, and digestive disorders
Sceletium tortuosum: Aizoaceae;	Traditionally used to relieve pain and alleviate hunger. May elevate mood and decrease anxiety, stress, and tension. Some report that it; it may potentiate the effects of Cannabis.
Schizozygia caffaeoides: Apocyanaceae;	Local medicinal uses include as a treatment for eye inflammation, sores, and ringworm-infected skin
Senegalia polyacantha subsp. *Campylacantha:* Fabaceae;	S. polycantha's root extract is useful for snakebites and is applied to wash the skin of children who are agitated at nighttime
Senegalia Senegal: Fabaceae	It is used as for its astringent properties, to treat bleeding, bronchitis, diarrhea, gonorrhea, leprosy, typhoid fever, and upper respiratory tract infections
Senna italic; Fabaceae:	Mild laxative
Solanum aculeastrum: Solanaceae	Traditional Zulu practitioners use the fruit - fresh, boiled, or charred - in herbal medicine to treat a wide variety of afflictions, including cancer, toothaches, and ringworm.

Continued...

Solanum americanum: Solanaceae;	Extracts were found to have selective antiviral activity against the herpes simplex type-1 virus (HSV-1). Methanol extracts of S. americanum have high antimicrobial activity against *Escherichia coli, Pseudomonas aeruginosa, Staphylococcus aureus* and *Aspergillus niger*. Water based extracts had no antibacterial activity
Solenostemma sp: Apocyanaceae;	The leaves are infused to treat gastro-intestinal cramps, stomachache, colic, and cold and urinary tract infections and are effective as an anti-syphilitic if used for prolonged periods of 40–80 days. Leaves have purgative properties that may be due to the latex present in the stems. Several active compounds have been extracted from Solenostemma. The native Sudanese have commonly used *Solenostemma argel* to suppress stomach pain, pains due to childbirth, and loss of appetite.
Spondias mombin: Anacardiaceae;	The fruit-juice is used as a febrifuge and diuretic. The roots are well-known febrifuge on the Ivory Coast, being sometimes used with leaves of X*imenia, Premna hispida, Ficus sp.,* and *Alchornea*
Strophanthus amboensis: Apocyanaceae;	S. *amboensis* is used in local medicinal treatments for rheumatism, venereal diseases, and scabies. The plant has been used as arrow poison.
Strophanthus boivinii: Apocyanaceae;	S. *boivinii* is used in local medicinal treatments for gonorrhea, colic, wounds, and itches
Strophanthus courmontii: Apocyanaceae,	S. *courmontii* is used in local medicinal treatments for rheumatism and as an aphrodisiac
Strophanthus eminii, S. eminii: Apocyanaceae,	is used in local medicinal treatments for snakebites, skin diseases and wounds and as an anthelmintic
Strophanthus preussii: Apocyanaceae;	Medicinal uses of S. *preussii* include treatment of gonorrhea and healing of sores. The plant has also been used as arrow poison.
Strophanthus sarmentosus: Apocyanaceae,	The many local medicinal uses of S. *sarmentosus* include treatment of joint pain, head lice, eye conditions and venereal disease. The plant has also been used as arrow poison
Strophanthus welwitschii: Apocyanaceae:	local medicinal treatments for respiratory conditions, gonorrhea, and scabies
Tabernamontana caffaeoides: Apocyanaceae,	Local medicinal uses include for weight loss and to combat fatigue.
Tabernaemontana crassa: Apocyanaceae	uses include anesthetic, hemostatic, and anthelmintic and in the treatment of rheumatism, kidney problems, rickets, and conjunctivitis. It has also been used as arrow poison

Tabernaemontana elegans: Apocyanaceae,	Its many local medicinal uses include the treatment of heart disease, cancer, tuberculosis, and venereal diseases. *T. elegans* is also used as an aphrodisiac.
Tabernaemontana pachysiphon: Apocyanaceae	Its many local medicinal uses include as a styptic, and as a treatment for headache, and to relieve cramps, hypertension
Tabernaemontana ventricosa: Apocyanaceae;	Local medicinal uses include the treatment of wounds, fever, and hypertension.
Terminalia schimperiana: Combretaceae:	The bark is applied to wounds, and the twigs may be chewed to promote oral hygiene. In laboratory experiments, extracts of the plant were found to have *in vitro* antibiotic properties against *Staphylococcus*. The plant extracts also have antifungal properties in vitro.
Thonningia: Balanophoraceae	A traditional remedy in many African cultures. This includes sexually transmitted diseases in Ghana, diarrhea in the Congo. In Zaire, it is said to prevent incontinence and bedwetting
Toddalia: Rutaceae:	The plant is used medicinally by many African peoples, including the Maasai, who use it for malaria, cough, and influenza. The roots have coumarins that have anti-plasmodial activity. Extracts of the plant have proved antiviral activity against H1N1 influenza in the laboratory.
Tulbaghia violacea: Amaryllidaceae;	Androgenic and anti-cancer properties *in vitro*. It also showed antithrombotic activities, which were higher than those found in garlic were.
Urtica massaica: Urticaceae	It is used in Rwanda to treat diarrhea. The Maasai use it to treat stomachache. They are used in Kenya to treat malaria. Other medicinal uses include treatment of fractures and venereal diseases
Vachellia karroo: Fabaceae	The gum, bark and leaves have been used as a soothing agent and astringent for colds, conjunctivitis, and hemorrhage in South Africa
Vachellia nilotica subsp. *Nilotica*: Fabaceae:	The bark is used to treat cough by the African Zulu
Vachellia sieberiana: Fabaceae:	In Africa, the bark or root is used to treat urinary tract inflammation. The bark has astringent properties and it is used to treat colds, cough, and childhood fever. According to the World Agroforestry Centre
Viscum capense: Santalaceae:	a tea made from plant parts is used to soothe asthma

Continued...

Voacanga thouarsii: Apocynaceae,	wounds, sores, gonorrhea, eczema, heart problems, hypertension, rheumatism, stomach-ache, and snakebite
Wrightia demartiniana: Apocynaceae	Treatment of kidney problems, gonorrhea and as a laxative.
Xylopia aethiopica: Annonaceae	An infusion of the plant's bark or fruit has been useful in the treatment of Asthma, Bronchitis, and dysentery, or as a mouthwash to treat toothaches. It has also been used as a medicine for biliousness and febrile pains.

Comments: Although there is some overlap between medicinal plants used in Africa and in Chinese, Indian and other medicinal systems, there are many plants that are unique to African folk medicine. More investigations are needed to find the chemical components that have medicinal properties.

Australian Aboriginal Traditional Medicine: Australia is the home of several unique plants that have grown and adapted to the harsh environment prevailing in much of Australia. Australian aboriginals have depended on and continue to depend on several native plants that they believe have medicinal properties. Aboriginal remedies varied between clans and in different parts of the country. There was no single set of Aboriginal medicines and remedies just as there was no one Aboriginal language. Aboriginal people used a range of remedies – wild herbs, animal products, steam baths, clay pits, charcoal and mud massages, string amulets, secret chants, and religious ceremonies to cure diseases. The aboriginal healers placed a great deal of emphasis on the supernatural as the cause of disease whereby all diseases were thought to emanate from a malfunctioning of the spirit located in the human body and so the healers treated patients by using plant products to take out the bad spirit and rejuvenate the body by putting back the good spirit. Throughout Australia, Aboriginal people believed that serious illness and death were caused by spirits or persons practicing sorcery. Even trivial ailments or accidents such as falling from a tree were often attributed to malevolence. Plants featured prominently in the treatment protocols of the Australian natives. Drinks made by steeping plant parts, body washes using plant decoctions, poultices, powders, and pastes of plant products as well as aromatherapy were some of the methods used to treat ailments. Since much of the information about the use of plants for medicinal purposes by native Australians is scattered, an online bioinformatics database known as the Customary Medicinal Knowledge base (CMKb) has been setup. http://biolinfo.org/cmkb/index.php. CMKb is an online relational database for collating, issuing, visualizing, and analyzing public domain data on customary medicinal plants. The database stores information related to taxonomy, photochemistry, biogeography, and biological activities of customary medicinal plant species as well as images of individual species. Known bioactive molecules are characterized within the chemo-informatics module of CMKb with functions available for molecular editing and visualization. All plants with potential medicinal values can be browsed for information and research. Over 450 such plants have been listed but most of them have not been characterized with respect to chemical constituents or bioactive molecules. The following is a short list of plants used by various aboriginal clans for medicinal purposes. More information can be found in the CMKb database and other sources. Here is a list of some native plants from the Aboriginal Pharmacopia (Crib, 1981).

Table 6: List of Medicinal plants in the Australian Aboriginal Pharmacopia

Botanical and Common name	Part used	Therapy
Abarema grandiflora – gins lips	bark	aphrodisiac
Acacia (Wattle)	Bark, gum, roots	Coughs and colds
Arpitania petriei (White Ash)	leaves, bark	Body pains
Balanophora fungosa (drumsticks)	shoot	aphrodisiac
Barringtonia racemosa (Yakooro)	Bark	Rheumatism, Sore throat
Basilicum polystachyon (Musk Basil)	Leaves	Fevers
Breynia stipitata	Leaves	Eye inflammation
Canavalia rosea (Beach bean)	Root infusion	Rheumatism, leprosy
Chamaesyce hirta (asthma plant)	Plant decoction tea	Asthma
Convolvulus erubescens (Bindweed)	Whole plant	Diahorrea, stomachaches
Cyathea australis (Rough tree fern)	Fronds	Tonic
Denhamia obscura	Inner bark	aphrodisiac, toothache
Dioscorea transversa (Native Yam)	Rhizome	Skin cancer
Eremophila gilesii (Turkey Bush)	Plant	Colds, body sores

Continued...

Eucalyptus gummifera (Red bloodwood)	Leaves	ringworm, venereal diseases
Eucalyptus microtheca (Coolibah)	Leaves, bark	Snake bite
Eucalyptus mannifera (Brittle Gum)	Gum	Laxative
Eucalyptus tetradonta (Darwin stringy bark)	Leaves, bark	Fevers, headache
Eucalyptus globulus (Tasmanian blue gum)	Leaves, gum	Asthma, colds, pain
Eudia vitiflora (Toothache tree)	Bark	Toothache
Euphorbia peplus (Canver weed)	Latex	Warts, skin cancer
Ficus (Native Fig)	Latex	ringworm,
Gratiola peruviana (Brooklime)	Leaves	Laxative, dizziness
Grewia retusifolia (Dysentery plant)	Leaves	Dysentery, diahorrea
Ipomoea pes-caprae (Coast morning glory)	Leaves	Hemorrhoids, diuretic, laxative
Leichardt australis (Australian Doubah)	Seeds	Oral contraceptive
Mentha satureioides (Native pennyroyal)	Leaves, oil	emmenagogue, Abortifacient
Myoporum platycarpum (Sugarwood)	Sugary sap	Mild laxative
Nauclea orientalis (Leichhardt tree)	Bark	Sore belly, possible anti-malarial
Sida rhombifolia (Paddy's Lucerne)	Young shoots	Diahorrea

Greco-Roman and European Medicinal Plants: Medicine in Europe had its recorded beginnings from the Greek philosopher scientist statesmen Hippocrates 460 BC-370BC, Aristotle 384 Bc-322 BC, Plato 428 BC-327 BC, Pliny 23-79 CE, Galen 129-200 CE and Dioscorides 40-90 CE. Of these, Hippocrates is widely considered the "Father of Medicine". Hippocrates recognized that human diseases are not caused by spiritual reasons but by physical imbalances caused by diet, environment, and physical causes. Hippocratic philosophy of medicine revolved on treatment rather than on diagnosis, that was the basis of the Knidian School of medicine." Corpus Hippocraticum Hippocratic corpus" is a collection of about 70 volumes that was written by Hippocrates himself and students of Hippocrates. The

Hippocratic School is credited with the description of various medical symptoms and conditions including clubbing of the fingers, suppuration of the lining of chest cavity, rectal examination, hemorrhoids, and conditions of the lungs. Hippocratic medicine was essentially based on the idea that the human body is composed of four humors namely black bile, yellow bile, phlegm and blood and the imbalance of any of these would affect the health of the individual. This theory is like the idea that the body is composed of Earth, Fire, Water, and air. As such, these ideas are also akin to the basis of Ayurveda, Siddha and Unani, Chinese and Islamic medical theories. Although modern medicine has disproved these ideas, many forms of alternative medicine continue to base their ideas on these ancient theories. Pliny wrote about the medicinal properties of plants in his "Historia Naturalis" Natural History. After Hippocrates and Pliny, the next significant physician was Galen a Greek who lived from AD 129 to AD 200. Galen expanded on the Humors theory initially advanced by Hippocrates and connected the role of each humor to human temperament. Thus, blood was connected to sanguine temperament, black bile with melancholic nature, yellow bile with a choleric temperament and phlegm with phlegmatic behavior. Individuals with sanguine temperaments are extroverted and social. Choleric people have energy, passion, and charisma. Melancholiacs are creative kind and considerate. Phlegmatic temperaments are characterized by dependability, kindness, and affection.

All the different schools of Greco-Roman medicine believed in a holistic form of medicine wherein plants played a big role as sources of drugs for treatment of diseases and for maintenance of good health. Medicinal plants like fennel, garlic, willow bark, Cannabis, Cumin, Aconite, Dill and other herbs were believed to help in curing diseases by bringing back the imbalance of the four humors during disease to a balanced situation so that the human body was brought back to good health.

Greco-Roman medicine recognized the plants listed below as having medicinal value and the medical practitioners used various parts of these plants to prepare poultices, decoctions, and powders. Very often, these medicinal plant extracts were steeped in grape wine and or olive oil before they were used for treatment.

The following is a short list of Greco-Roman, Iranian, and Egyptian medicinal plants. Many other herbals used in Indian and Chinese traditional medicine are also components of this system and so are not listed.

Aconite (*Aconitum napellus*), also known as Monkshood Wolf's Bane, Aniseed or (*Anise Pimpinella*), Artemisia/Worm wood (*Artemisia absinthium*), Artichoke (*cynara scolymus*):, Asparagus (*Asparagus officinalis*), Bluebell (*Hyacinthus nonscriptus*), Cabbage (*Brassica oleracea*), Cardomum (*Eletteria cardamomum*), Chamomile (*Matricaria Chamomilla* and *Anthemis nobilis*), Marijuana/hemp (*Cannabis* sp.), Chestnut (*Castanea sativa*), Chicory (*Cichorium intybus*), Cherry (*Prunus avium*), Cinnamon (*Cinnamum zeylanicum*), Citrus fruits such as Oranges and Lemons Centaury (*Erythraea centaurium*), Cumin: (*Cuminum siminum*), Dog's Mercury (*Mercurialis perennis*), Elecampane (*Inula helenium*), Eyebright (*Euphrasia officinalis*), Fennel (*Foeniculum vulgare*), Fenugreek (*Trigonella foenumgraecurn*):, Fig – (*Ficus carica*), Flax (*Linum usitassimum*) Garlic (*Allium sativum*), Horehound (*Marrubium vulgare*), Hyssop (*Hyssopus officinalis*), Juniper *Juniperus sp.*, Lettuce (*Lactuca scariola*), Mallow (*Malva sylvestris*), Mint – (*Mentha piperata, Mentha pulegium*). Mistletoe (Loranthus), Mulberry (*Morus nigra*), Oregano (*Origanum vulgare*), Myrtle (*Myrtus communis*), Narcissus (*Narcissus poeticus*), Nightshade (*Atropa belladonna*), Oak (*Quercus*), Parsley (*Petroselinum crispum*), Plantain (*Plantago major, P. Minor, P. lanceolate)*, Pomegranate (*Punica granatum*), Poppy (*Papaver somniferum*), Quince (*Cydonia vulgaris*), Rice (*Oryza sativa*), Rose (*Rosa* sp): Rosemary (*Rosmarinus officinalis*), Rue (*Ruta graveolens*), Savory, Winter savory – (*Satureja Montana*), Summer savory – (*S. hortensis*), Stinging Nettle (*Utrica dioica*), Tarragon (*Artemisia dracunculus*), Thyme – (*Thymus vulgaris*), Uva Ursi (*Arctostaphylos uva ursi*), Vine Grape *Vitis vinifera*, Violets (*viola odorata*), Walnut (*Juglans nigea)*, Winter savory (*Satureja montana*), Yarrow (*Achillea millefolium*).

Latin America and Amazonian forests: This region covering Mexico, Central America and South America is a vast region, which is rich in many food and medicinal plants. The Aztecs, Incas and Mayan cultures that ruled in this region had developed their own individual systems based on combination of spiritual incantations and use of plants to treat ailments. In general, emphasis was placed on curing the mind along with the body and the spirit there and treatments were verbally transferred within families and within regions. However, after the Spanish conquest of the region medical information was spread among larger regions. The folk herbal medicine in this region was based on the plants that grew in the region particularly in the rainforests and in the high elevations of the Andes. Many plants particularly belonging to the Solanaceae such as Chili peppers, Potato and Tomato are natives of the Americas and were imported to the old-world countries of Europe and Asia. Other major plants include Corn and several other medicinals such as Cinchona-Quinine, coca, curare, cacao, sumac, Guarana, Lapachol, Stevia and Ipecac, The list of plants below is based on traditional usage by the populations in Latin America particularly in the Amazon region. Very few of these have been examined by modern methods to verify their claims.

Acai Berry (*Euterpe oleracea*): The berry is said to have high levels of antioxidants and is now extremely popular in health food stores.

Albahaca or Basil (*Ocimum micranthum*): Headaches, migraine, kidney and bladder problems, dizziness, neuritis, and cardiovascular diseases.

Arrowroot powder (*Maranta arundinacea*): Used as infant food and for control of diarrhea.

Boldo (*Pneumus boldus*): Boldo leaves have been used by indigenous peoples in Chile and Peru for liver ailments and in the treatment of gallstones.

Branco/Huilco, Huilca, Wilco (*Anadenanthera colubrine*): This deciduous tree bark is astringent, bitter, purgative, and hemostatic. A kind of snuff made of bark powders is used to treat respiratory diseases.

Cacao (*Theobroma cacao*): The Aztec name for this tree is chócolatl from which comes the English word chocolate. Cacao seed pulp was used to stimulate the nervous system. Cocoa bean powder and dark chocolates have high antioxidant properties and hence recommended for treating plaque formation in blood vessals.

Cat's Claw (*Uncaria tomentosa*): Immunostimulant and anti-inflammatory agent.

Catuaba (*Erythroxylum catuaba*): Known as caramuru, catagu, catigu, catigua, chuchuhuasha, pau de reposta or tatuaba is a Brazilian aphrodisiac with properties for treating erectile dysfunction.

Chanca Piedra (*Phyllanthus niruri, Phyllanthus, quebra*): kidney stone tree. This plant is also prescribed by Ayurveda, Unani, and Siddha forms of medicine for liver and kidney ailments.

Chilean Holly/Taique, (*Desfontainia spinosa* var. hookeri): The leaves and fruits are narcotics used by the Mapuche people of Chile. The leaf and fruit extracts are used by Shamans to divine visions because of its psychoactive powers.

Chuchuhuasi (*Maytenus krukovii*): This is an amazon rainforest tree whose bark extracts have phytomedicines that improve back pains, arthritis, and rheumatism.

Cinchona/Quina/Quechua (*Cinchona officianalis, C. calisaya,* **and** *Cinchona pubescens*): These species of Cinchona especially *C. Officianalis* are highly valued medicinal plants of this region. These trees have been introduced into India, Indonesia, and other countries. The Cinchona bark is the source for quinine that is a powerful anti-malarial alkaloid. Although synthetic quinine and quinine-based drugs have replaced cinchona, the tree bark still is a major source for treating malaria in ethno botanical medicines.

Coca (*Erythroxylum coca*): It is the source of cocaine and the leaves, bark and other plant parts are chewed by the tribes of South and Central America as a stimulant and for control of pain.

Common Nasturtium (*Tropaeolum majus*): In local traditional medicine, plant extracts are used as a remedy for cough, cold, asthma, diabetes, anemia, constipation and as a body deodorizer. In Peru, this plant is commonly used to improve the healing of wounds. Flower and leaf extracts have anti-bacterial properties.

Contrayerva, or contrajerva, (*Dorstenia* sp.): This plant growing in the Columbian region has hallucinogenic properties.

Copaibo (*Copaifera officinalis L.*): This plant is known variously as Palo de aceite, Oily stick, Spanish Raë. The Tree is well known to the inhabitants of the Amazonian region for its medicinal properties. It is used as a treatment for colic, rheumatism, and kidney cleansing.

Curare (*Chondodendron tomentosum*): grows wild in the Amazonian rain forest. The roots and rhizomes have a poisonous alkaloid namely Tubocurarine which is used during surgery to paralyze muscles.

Damiana (*Turnera diffusa*): Damiana is used primarily as an aphrodisiac for both sexes but its folklore uses also extend to include treatment for depression and anxiety.

Guayaba/Guava: (*Psidium guajava L.*): The fruit has laxative properties. The infusion of Guayabo flowers is a traditional herbal remedy for sexual weakness, foul smelling vaginal discharges and severe abdominal cramps. The tincture of Guayabo leaves is a particularly good remedy for the treatment of varicose veins. The infusion of the bark of Guayabo is a traditional remedy for the treatment of psoriasis and heat rashes.

Epazote (*Chenopodium ambrosioides*): This plant is known as Wormseed, Jesuit's Tea, Mexican Tea and *Herba Sancti Mariæ*. The use of epazote can be traced back to the Aztecs. Epazote has up to 70% Ascaridole and is a treatment for intestinal round worms.

Guarana (*Paullinia cupana*): The seeds have more caffeine than coffee and are a stimulant.

Hercampuri (*Gentianella alborosea*): This plant is a species in the family Gentianaceae. Plant infusions are used in traditional medicine for treating hepatitis, varicose veins, reduction of blood cholesterol and hypertension.

Horsetail (*Equisetum giganteum* L): It is known in the Amazonian regions of Bolivia for the treatment of tuberculosis infections of the gums and oral mucosa.

Ipecac (*Cephaelis ipecacuanha*): The leaves and bark of this plant is used by South Americans to clear the stomach and respiratory tract cough and colds. European explorers took it home with them in 1672 and found it to be an effective treatment for amoebic dysentery. It is also an emetic.

Iporuru/Iporoni (*Alchornea castaneifolia*): Shrubby tree native to Brazilian rain forest. It is recommended highly for treating arthritis and rheumatism

Jaborandi (*Pilocarpus annatifolius*): This native of Central and South America is a diaphoretic. Diaphoretics stimulate profuse sweating usually through the action of glycosides. The drug Pilocarpine is derived from this plant.

Jamaican Dogwood (*Piscidia piscipula*): This plant is commonly known as Florida Fishposion Tree, Jamaican dogwood tree. It is well known as a specific for migraine headaches and neuralgia and for treatment of insomnia caused by muscle tension and stress.

Karé (*Chenopodium ambrosioides* L): Seeds are used for Intestinal bloating and colitis.

Lapacho (*Tabebuia avellandedae* and *T. impetiginosa.*): It is also commonly known as Pau D'arco. The bark has Lepachol, which in rats has been shown to have abortifacient effects. Bark extracts have anti-viral, anti-fungal properties. It has anti-blood coagulant properties. It was considered as a treatment for cancers but abandoned due to its extreme toxicity.

Muira Puama (*Liriosma ovata*): A native folk medicine of Brazil muria puama is best known as an aphrodisiac.

Mulungu (*Erythrina mulungu*): Tinctures and decoctions of bark of this tree are used as fish poisons as well as sedative.

Muña (*Minthostachys mollis*): This is a medicinal plant restricted to the South American Andes from Venezuela to Bolivia. Plant roots are boiled, and the decoction is drunk to cure body pains.

Papaya (*Carica papaya*): The fruits are rich in papain and chymotrypsin and are used as a digestive aid.

Passionflower (*Passiflora incarnate*): It has mild sedative effects and used in sleep preparations.

San Pedro Cactus (*Echinopsis pachanoi*) **and Peruvian torch cactus** (*E. peruviana*): Both cacti found in Argentina, Bolivia, Chile, Ecuador, and Peru have the psychoactive alkaloid mescaline, which is a psychedelic alkaloid.

Snakeweed/Cayenne (*Stachytarpheta cayenensis*): Plant extracts have anti-inflammatory, anti-ulcerous, anti-diarrheal, antibacterial, antiviral, antispasmodic properties. The plants also grow in many Asian countries. They are used in traditional native medicines. The leaves hold cyanides and hence extracts must be boiled before consumption.

Stevia (*Stevia rebaudiana*): The leaf of the stevia is several hundred times sweeter than sugar and is calorie-free. It is fast becoming one of the most important sugar substitutes.

Suma (*Pffafia paniculata*): Claimed to be an aphrodisiac, tonic energizer with anti-cancer properties

Tooth ache plant: (*Acmella oleracea*); This Brazilian rain forest plant having analgesic alkyl amides called spilanthol is considered a natural muscle relaxant and has been traditionally used as an herbal to numb toothaches.

Wild Yam Root (*Dioscorea villosa*) Wild yam root is a potent source of diosgenin that was used in the manufacture of the first oral contraceptives.

Yacon Root (*Smallanthus sonchifolius*): Yacon root and yacon root syrup are used as a sweetener.

Yerba mate (*Ilex paraguariensis*): This member of the holly family is used as a stimulant like coffee.

American Indian Folk Medicine: The American Indian tribes used plants along with spiritual healing to treat diseases because they believed that various evil spirits oversaw diseases. Plants of medicinal use among this population consisting of various tribes like the Chippewa, Comanche, Pawnee, Shiawassee, Mohawk, Menominee, Navaho, Shoshoni, yopkia, are listed below with their presumed value for treating diseases. Although some of these are marketed in health supplements, both toxicity and side effects must be examined before considering usage.

Abdominal Pain/Cramps: Cattail (Typha); Sage (*Salvia officinalis* L.); Saw Palmetto (*Serenoa repens*).

Abortifacient: Pennyroyal (*Hedeoma pulegioides*); Rosemary (*Rosmarinus officinalis*); Skullcap (*Scutellaria*); Slippery Elm (*Ulmus rubra*).

Abscesses: Devil's Claw (*Echinacea*); Chamomile (*Matricaria chamomilla*); Pau d'arco (*Tabebuia cassinoides*); Slippery Elm (*Ulmus rubra*) Wild Yam (*Dioscorea villosa*).

Aches: Black Cohosh (*Actaea racemosa*); Devil's Claw (*Echinacea*).

Analgesic, antiseptic: Yarrow (*Achillea millefolium*); Arnica (Arnica sp.); Horsemint (*Monarda fistulosa*), Wintergreen (*Gaultheria procumbens*).

Antirheumatic: Willow Bark (*Salix*).

Anemia: Senna (*Cassia Senna*), Wheat Grass (*Triticum sp*).

Blood Remedy: Sassafras (*Sassafras albidum*).

Boils: Grape (*Vitis vinifera*), Peach Leaves (*Prunus sp.*)

Colds coughs and respiratory problems: Putty root (*Aplectrum hymale*); Indian tobacco (*Lobelia inflate*); Maiden hair fern (*Adiantum Capillus-veneris*); Aaron's rod (*Verbascum Thapsus*), Dogwood, Willow (*Salix sp*).

Diuretic: Star Grass (*Aletris farinose*): Burdock (*Arctium lappa*); Yellow parilla (*Menispermum candense*); Butterfly weed (*Asclepia tuberosa*).

Gastro Intestinal problems/Stomach Ache: Dandelion (*Taraxacum officinale*); Fennel (*Foeniculum vulgare*); Rabbit Tobacco (*Pseudognaphalium obtusifolium*); Sage (*Salvia officianalis*); Saw Palmetto (*Serenoa repens*); Star Anise (*Illicium verum*); Wild Garlic (*Allium canadense*); Golden Seal (*Hydrastis canadensis*).

Heart and Circulatory Problems: Dogbane (*Apocynum cannabinum*).

Inflammations and Swellings: Witch Hazel (*Hamamelis virginiana*),

Insect Repellents and Insecticides: Goldenseal (*Hydrastis canadensis*).

Laxative: Curly dock (*Rumex crispus*).

Rheumatism: Pokeweed (*Phytolacca Americana*); Bloodroot (*Sanguinaria canadensis*); Black willow (*Salix nigra*). Warts: Milk Weed (*Asclepias syriaca*).

Polynesia

This region consisting of over 1000 islands in the pacific spanned the islands of New Zealand, Maori, Fiji, Samoa, Tonga, and the Cook Islands, Hawaii, Tahiti, Marshall Islands, Micronesia, and several others. As in other cultures, plant-based medicines play a large role. Some of the most prominent medicinal plants in this region are listed below.

1. **Kava Kava–** (*Piper Methysticum*): Pain-relieving properties that can reduce headaches and back pains drunk as a beverage. It has muscle relaxant properties. 2. **Mamaki** – (*Pipturus Albidus*) tea-like beverage used to relive cough, sore throat, and thrush. 3. **Noni** (*Morinda Citrifolia*), curing severe ailments like high blood pressure, heart diseases, and diabetes. 4. **Olena/Turmeric** (*Curcuma longa*), treatment for inflammation, skin ailments, sore throat. 5. **Kukui/Candlenut** (*Aleurites Moluccana*), Skin sores.

Other medicinal herbs are:

A'atasi (*Rorippa sarmentosa* (DC.). Leaves for carbuncle to promote healing of wounds- Juice instilled into eyes.

Aloalo (*Premna serratifolia* L.) used externally as a poultice for sores and wounds.

'Aoa (*Ficus obliqua Forst.*): Leaves used externally for carbuncle

Apu initia (*Anacardium occidentale*): Fruit taken internally for sore throat and difficulty in swallowing.

Ateate (*Wollastonia biflora*): Leaves taken internally for enlarged liver, inner bark taken internally for urinary tract infections.

Au'auli (*Diospyros samoensis*): Leaves taken internally for hypertension.

Ava'ava aitu (*Macropiper puberlum*): Leaves and inner bark taken internally for erythremia and cellulitis.

'Ava niukini (*Derris malaccensis* L.): Pounded root used as a poultice for ringworm and lice.

Ava pui vao (*Zingiber zerumbet*): Roots, taken internally for stomachache

Fetau/Alexandrian (*Calophyllum inophyllum*): Leaves broken in container of seawater to bathe skin rash or infections.

Filimoto (*Flacourtia ruka*): Inner bark taken internally for septicemia.

Fu'afu'a (Kleinhovia hospita L): Inner bark for bloody stool in children.

Fue moa *(Ipomoea pes-caprae.L.)*: Leaves used externally to treat chicken pox.

Ku'ava (*Psidium guajava* L): Inner bark extracts taken internally for intestinal tract diseases of children and for treating diphtheria, and sore throat.

Lau gasese- Samoan name (*Davallia solida*): Leaves used externally as poultice for arthritis.

Lau tamatama (*Achyranthes aspera* L.): Leaves used as a poultice for carbuncles and abscesses, Leaves used to promote healing of wounds, Leaves taken internally as a general childhood tonic.

Leva (*Cerbera manghas* L): Roots used as a treatment for cancer.

Magele (*Trema cannabina*): Root powders and extracts taken internally for diarrhea, skin rash, and hypertension and as a laxative.

Moso'oi/Perfume tree (*Cananga odorata*): Inner bark as a laxative

Pua Samoa/Tahitian gardenia (*Gardenia taitensis* DC.): Leaves taken internally for diabetes, used for "cleansing the blood".

Talie/Tropical almond (*Terminalia catappa* L.): Inner bark extracts taken internally for childhood diseases of the intestinal tract, and diphtheria.

Togo/Asiatic pennywort (*Centella asiatica* (L.): Leaves used internally and externally for mumu afi Leaves used together with fue sina internally and externally for pua'i toto.

Kawakawa/)/Pepper tree. (*Macropiper Excelsum*): Externally used for cuts, wounds, bruises and rheumatism and for the pain of neuralgic conditions, toothache, nettle stings, eczema, venereal diseases and festering sores The leaves were often chewed and used for stomach pains and indigestion. Kawakawa also stimulates the appetite. Diuretic, anti-inflammatory and anti-neuralgic activities. Internally, used as a blood purifier and tonic.

Manuka/tea tree (*Leptospermum scoparium*) The Manuka has been traditionally used by the New Zealand Maori for a variety of complaints internally and externally. Vapor baths were used to treat lumbago, rheumatism, and ease childbirth. Leaves, seeds, and bark after preparation were also used as poultices and ointments for ringworm, burns and scalds, sprains, wounds, lesions from venereal diseases and eczema.

References

http://nischennai.org/siddhamedicine.html

https://www.medicinenet.com/script/main/art.asp?articlekey=20281

http://www.ccras.nic.in

http://apps.who.int/medicinedocs/pdf/s7916e/s7916e.pdf

http://tkdl.res.in/tkdl/langdefault/Unani/Una_ccrum.asp?GL=Eng

https://www.ctahr.hawaii.edu/adap/publications/adap_pubs/1993-1.pdf

http://ayush.gov.in/about-the-systems/siddha

http://ayush.gov.in/about-the-systems/ayurveda

http://ayush.gov.in/about-the-systems/unani

https://www.atms.com.au/modalities/herbal-medicine

https://www.westernsydney.edu.au/nicm/health_information/links

http://www.cintcm.ac.cn/opencms/opencms/en/database.html

http://www.tcmbasics.com/materiamedica.htm

http://www.gateway-africa.com/tradmedicine

https://www.springer.com/us/book/9783540419297

Books and Journals

Achyutha warrier, C., (1991): Ashtangahrudayam, Uttarasthanam, (Malayalam), Devi Book Store, Kodungalloor, India.

Achyutha warrier, C., (1995): Ashtangahrudayam, Soothra sthanam (Malayalam), Devi Book Store, Kodungalloor, India.

Barr, A., Chapman, J., Smith, N. & Beveridge, M. (1988): Traditional Bush Medicines, an Aboriginal Pharmacopoeia. Melbourne: Greenhouse Publications

Bhushan, P., Kalpana, J. & Arvind, C, (2005): Classification of human population based on HLA gene polymorphism and the concept of Prakriti in Ayurveda. J. Altern. Complement Med. 11, 349–353.

Castro, Luisa, (2001): Samoan Medicinal Plants and Their Usage 2001 edition edited and designed by University of Hawaii at Manoa Kristie Tsuda, University of Hawaii at Manoa ISBN 1-931435-27-8.

Clarke, P., (2008): Aboriginal healing practices and Australian bush medicine, Journal of the Anthropological Society of South Australia Vol. 33 – 2008, 1-25

Cribb, A.B., and J.W. Cribb (1981), Wild medicine in Australia, William Collins Pvt Ltd, Sydney, pp228.

Dan Bensky, Ted J. Kaptchuk (1993): Chinese Herbal Medicine: Materia Medica, Eastland Press, 1993 - Science - 556 pages

Dwarakanath, C. (1952): The Fundamental Principles of Ayurveda. (Krishnadas Academy, Varanasi, India,

Fawzi Mahomoodally, M.F., (2013): Traditional Medicines in Africa: An Appraisal of Ten Potent African Medicinal Plants, Evidence-Based

Complementary and Alternative Medicine, Volume 2013, Article ID 617459, 14 pages. http://dx.doi.org/10.1155/2013/617459, Review Article.

Govindaraj,P.; Nizamuddin,S., Sharath,A., Jyothi,V., Rotti, H., Raval,R., Nayak,J., Bhat,B.K., Prasanna, B. V., Shintre, P., Sule,M.,. Joshi, K.S., Dedge, A.P., Bharadwaj, R., Gangadharan, G. G., Nair, S., Gopinath, P.M., Patwardhan, B., Kondaiah, P., Satyamoorthy, K., Valiathan, M.V.S., & Thangaraj, K. (2015): Genome-wide analysis correlates Ayurveda Prakriti. Sci. Rep. 5, 15786; doi: 10.1038/srep15786 (2015).

Hankey, A. (2001): Ayurvedic physiology and etiology: Ayurvedo Amritanaam. The doshas and their functioning in terms of contemporary biology and physical chemistry. J. Altern. Complement Med. 7, 567–574 (2001).

Hankey, A. (2005): A test of the systems analysis underlying the scientific theory of Ayurveda's Tridosha. J. Altern. Complement Med. 11, 385–390 (2005).

Jayasundar, R. (2010): Ayurveda: a distinctive approach to health and disease Curr. Sci., 98, 908–914 (2010).

Lassak, E.V. and T. McCarthy (2011): Australian medicinal plants: a complete guide to identification and usage/2nd ed.; 303 p., [20] p. of Chatswood, N.S.W.: Reed New Holland, 2011.

Mahalle, N. P., Kulkarni, M. V., Pendse, N. M. & Naik, S. S. (2012): Association of constitutional type of Ayurveda with cardiovascular risk factors, inflammatory markers, and insulin resistance. J. Ayurveda Integr. Med. 3, 150–157 (2012).

Maurice M. Iwu (2014): Handbook of African Medicinal Plants, CRC Press, Feb 4, 2014 - Health & Fitness - 506 pages.

Mazzari, Andrea L. D. A.; Prieto, Jose M. (2014). "Herbal medicines in Brazil: Pharmacokinetic profile and potential herb-drug interactions". Frontiers in Pharmacology. 5: 162. doi: 10.3389/fphar.2014.00162. PMC 4087670. PMID 25071580.

Padmasiri, G.R., (2018): An investigation into utilization, beliefs, and practice of indigenous medicine in Sri Lanka. African Journal of Traditional, Complementary and Alternative Medicines - Vol 15, No 4, p1-12.

Park, H.L., Lee, H.S., Shin, B.C., Liu, J.P., Shang, Q., Yamashita, H., et al. (2012): Traditional medicine in China, Korea, and Japan: a brief introduction and comparison, Evid Based Complement Alternate Med, 2012 (2012), p. 429103.

Patwardhan, B. (2003): AyuGenomics–Integration for customized medicine. Indian J. Nat. Prod. Resource. 19, 16–23.

Pyo, K. (1987): The significance of Dongje Medical School in medical history, J Korean Med Classic, 1 (1987), pp. 16-18.

Rotti, H. et al. (2014): Immuno-phenotyping of normal individuals classified based on human Dosha prakriti. J. Ayurveda Integr. Med. 5, 43–49 (2014).

CHAPTER 4

Human Genetics and Genetic Diseases

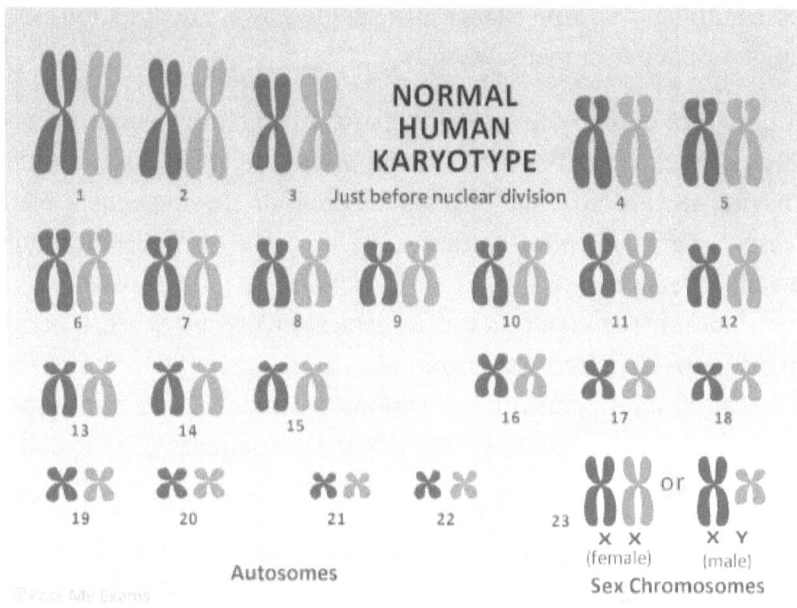

The bulk of genetic information in the human cell is packaged into chromosomes found in the nuclei. However, some genes particularly those involved in energy generation and muscle development are in small organelles called mitochondria. Defects in the chromosomal nuclear DNA as well as in the circular DNA of mitochondria can affect normal gene expression and thereby result in genetic abnormalities in the individuals carrying these defective genes.

The human cell nucleus has 22 pairs of autosomal chromosomes plus a pair of X chromosomes in females and 22 pairs of autosomal chromosomes plus 1X chromosome and 1Y chromosome in males. Autosomal chromosomes are those chromosomes that are not directly involved in sex determination. Chromosomes 1-22 are autosomal chromosomes. The X chromosomes have many genes including those involved in female determination while the Y chromosome that is a modified X-chromosome has very few genes, which are concerned with determination of maleness.

Genetic disorders can be caused by 1. Chromosomal aberrations due to duplications, deletions, inversions and translocations in the chromosomes 2.mutation in one gene (monogenic disorder), 3. mutations in multiple genes (multifactorial inheritance disorder), 3.by a combination of gene mutations and environmental factors.

1 Duplication of chromosomes or parts of a chromosome such that instead of 46 chromosomes the person has 47 or more chromosomes or 46+ having 46 chromosomes plus part of another chromosome. Deletions are cases where an entire chromosome or parts of a chromosome are removed. Inversions are those cases wherein the normal order of genes in a chromosome is changed and translocations result when a part of a chromosome is transferred on to another chromosome. Such chromosomal changes involving duplications, deletions, inversions, and translocations can result in abnormal gene expression and consequent genetic diseases. Examples are:

Duplications: Downs syndrome, 47+21, Patau syndrome 47+13, Edwards syndrome, 47+18; Kleinfelters syndrome 47XXY, 47XXX syndrome, 47XYY syndrome.

Deletions: Turner syndrome, 45, X; Cri-du-chat syndrome, 46-5p; Retinoblastoma, deletions in long arm of chromosome 13, deletions in this region also shown in some patients with acute myeloid leukemia, breast, bladder, lung and prostate cancers. The RB gene may be implicated in multiple cancers.

Inversions: Type 3 chromosome 15

Translocations; Downs's syndrome 14/21, Alagille syndrome 2/20, Burkitts lymphoma 8/14; chronic myeloid leukemia Philadelphia G chromosome 9/22; acute promyelocytic leukemia (Ch. 15/17)

Iso chromosomes and ring chromosomes: In many clinical cases, ring formations and isochromosomes had no apparent phenotypes but in some cases, they correlate with clinical manifestations as in Ch. 22 ring – cat eye syndrome, dicentric chromosome1 and chromosome 16 causes multiple disorders including mental retardation and developmental defects.

Monogenic disorders: There are many diseases that manifest when there is a mutation in a single gene. These mutations can be dominant or recessive. If dominant, then the condition is expressed even if the mutation occurs in only one of the two homologous chromosomes. However, if it is a recessive condition then such mutation is expressed only if both homologs carry the same mutation. If only one of the chromosomes has the recessive mutation (heterozygous) then, the person carrying the mutation becomes a carrier and passes it on to the progeny. If the mutation occurs on the X chromosome, then it will be expressed whether dominant or recessive in males since males have only one X chromosome. However, in females. X linked genetic diseases are expressed only if they are dominant or if both the X chromosomes carry the recessive mutation.

Autosomal genetic disease is those caused due to dominant mutations in any one of chromosomes 1 – 22 or recessive mutations caused on both pairs of any one of chromosomes 1 – 22. Examples of autosomal monogenic dominant diseases are achondroplasia, polycystic kidney disease, phenyl ketone urea. Examples of recessive monogenic mutations are cystic fibrosis, sickle cell anemia.

All the above will be passed on from parent to progeny through the sperm of the male or the egg of the female. There are also possibilities of mutations occurring in the somatic body cells of an individual due to errors occurring during replication of the DNA or due to exposure to radiation, chemicals, and other environmental causes. Such errors that occur in an individual will not be passed on to the progeny. Genetic diseases are not contagious.

Genetically inherited diseases cannot be prevented but can be diagnosed through amniocentesis and other DNA tests and the probability of inheriting a certain type of genetic disease can be predicted by examining the family tree of the concerned individual. While certain genetically inherited diseases can be detected at an early stage, others are expressed only at later stages in life. While the expression of many genetically inherited conditions cannot be prevented, their expression in some cases such as Phenyl ketone urea PKU can be minimized by proper nutritional control. Likewise, proper nutritional control and use of plant-based supplements can prevent/reduce and delay the onset of certain diseases such as diabetes and cardiovascular diseases and even the development and metastasis of cancers.

Autosomal dominant: Autoimmune lymphoproliferative syndrome, Arteriovenous malformation, Autoimmune lymphoproliferative syndrome, Capillary malformation-AV malformation syndrome, Cardiofaciocutaneous syndrome, Costello syndrome, Gingival fibromatosis with hypertrichosis, Hereditary gingival fibromatosis type 1, Legius syndrome, Moynahan syndrome, Neuro-cardio-facial-cutaneous syndromes, Neurofibromatosis type I, Noonan syndrome, Noonan syndrome with multiple lentigines, Noonan-like Legius syndrome, Acropectoral syndrome, Acute intermittent porphyria, Adermatoglyphia, Albright's hereditary osteodystrophy, Arakawa's syndrome II, Aromatase excess syndrome, Autosomal dominant cerebellar ataxia, Axenfeld syndrome.

Bainbridge-Ropers Syndrome, Benign hereditary chorea, Bethlem myopathy, Birt–Hogg–Dubé syndrome, Boomerang dysplasia, Branchio-oto-renal syndrome, Buschke–Ollendorff syndrome, Camurati–Engelmann disease, Central core disease, Collagen disease, Collagenopathy, types II and XI, Congenital distal spinal muscular atrophy, Congenital stromal corneal dystrophy, Costello syndrome, Currarino syndrome.

Darier's disease, Glut1 deficiency, Dentatorubral-pallidoluysian atrophy, Dermatopathia pigmentosa reticularis, Dysfibrinogenemia, Emberger syndrome, Familial amyloid polyneuropathy, Familial atrial fibrillation, Familial hypercholesterolemia, Familial male-limited precocious puberty, Feingold syndrome, Felty's syndrome, Flynn–Aird syndrome, Gardner's syndrome, GATA2 deficiency, Gillespie syndrome, Gray platelet syndrome, Greig cephalopolysyndactyly syndrome, Hagemoser–Weinstein–Bresnick syndrome, Hajdu–Cheney syndrome, Haploinsufficiency of A20, Hawkinsinuria, Hay–Wells syndrome, Helsmoortel-Van der Aa syndrome, Hereditary elliptocytosis, Hereditary hemorrhagic telangiectasia,

Hereditary mucoepithelial dysplasia, Hereditary spherocytosis, Holt-Oram syndrome, Huntington's disease, Huntington's disease-like syndrome, Hypertrophic, ardiomyopathy, Hypoalphalipoproteinemia, Hypochondroplasia, Hypodysfibrinogenemia, Jackson–Weiss syndrome.

Keratoendotheliitis fugax hereditaria, Keratolytic winter erythema, Kniest dysplasia, Kostmann syndrome, Langer–Giedion syndrome, Larsen syndrome, Liddle's syndrome, Marfan yndrome, Marshall syndrome, Medullary cystic kidney disease, Metachondromatosis, Miller–Dieker syndrome, MOMO syndrome, Monilethrix MonoMAC, Multiple endocrine neoplasia, Multiple endocrine neoplasia type 1, Multiple endocrine neoplasia type 2, Multiple endocrine neoplasia type 2b, Myelokathexis, Myotonic dystrophy, Naegeli–Franceschetti–Jadassohn syndrome, Nail–patella syndrome, Noonan syndrome, Oculopharyngeal muscular dystrophy, Pachyonychia congenital, Pallister–Hall syndrome, PAPA syndrome, Papillorenal syndrome, Parastremmatic dwarfism, Pelger-Huet anomaly, Peutz–Jeghers syndrome, Piebaldism, Platyspondylic lethal skeletal dysplasia, Torrance type, Polydactyly, Popliteal pterygium syndrome, Porphyria cutanea tarda, Pseudoachondroplasia, RASopathy (Capillary malformation-AV malformation syndrome, Autoimmune lymphoproliferative syndrome, Cardiofaciocutaneous syndrome, Hereditary gingival fibromatosis type 1, Neurofibromatosis type 1, Noonan syndrome, Costello syndrome, Noonan-like[1], Legius syndrome, Noonan-like Noonan syndrome with multiple lentigines, formerly called LEOPARD syndrome); Reis–Bucklers corneal dystrophy, Romano–Ward syndrome, Rosselli–Gulienetti syndrome, Roussy–Lévy syndrome, Rubinstein–Taybi syndrome, Saethre–Chotzen syndrome, Schmitt Gillenwater Kelly syndrome, Short QT syndrome, Singleton Merten syndrome, Spinal muscular atrophy with lower extremity predominance, Spinocerebellar ataxia, Spinocerebellar ataxia type 1, Spinocerebellar ataxia type 6, Spondyloepimetaphyseal dysplasia, Strudwick type, Spondyloepiphyseal dysplasia congenital, Spondyloperipheral dysplasia, Stickler syndrome, SYT1-associated neurodevelopmental disorder, Tietz syndrome, Timothy syndrome, Treacher Collins syndrome, Tricho–dento–osseous syndrome, Tuberous sclerosis, Upington disease, Variegate porphyria, Vitelliform macular dystrophy, Von Hippel–Lindau disease, Von Willebrand disease,

Wallis–Zieff–Goldblatt syndrome, WHIM syndrome, White sponge nevus, Worth syndrome, Zaspopathy, Zimmermann–Laband syndrome, Zori–Stalker–Williams syndrome.

Autosomal recessive

2-Hydroxyglutaric aciduria, 2-Methylbutyryl-CoA dehydrogenase deficiency, 3-Methylcrotonyl-CoA carboxylase deficiency, 3C syndrome, 6-Pyruvoyltetrahydropterin synthase deficiency, 17β-Hydroxysteroid dehydrogenase III deficiency, Abdallat–Davis–Farrage syndrome, Abderhalden–Kaufmann–Lignac syndrome, Abetalipoproteinemia, Ablepharon macrostomia syndrome, Acatalasia, Aceruloplasminemia, Acheiropodia, Acrocallosal syndrome, Acrodermatitis enteropathica, Acute fatty liver of pregnancy, Adducted thumb syndrome, Adenine phosphoribosyltransferase deficiency, Adenosine deaminase 2 deficiency, Adenosine deaminase deficiency, Adenylosuccinate lyase deficiency, Al-Raqad syndrome, Albinism in humans, Aldolase A deficiency, Alkaptonuria, ALOX12B, Alpha-aminoadipic and alpha-ketoadipic aciduria, Alpha-mannosidosis, Aminoacylase 1 deficiency, Aminolevulinic acid dehydratase deficiency porphyria, Antley–Bixler syndrome, Apparent mineralocorticoid excess syndrome, Arginine: glycine amidinotransferase deficiency, Argininemia, Argininosuccinic aciduria, Arterial tortuosity syndrome, Aspartylglucosaminuria, Atelosteogenesis, type II, Atransferrinemia, Autosomal recessive multiple epiphyseal dysplasia.

Baller–Gerold syndrome, Bare lymphocyte syndrome, Bare lymphocyte syndrome type II, Batten disease, Behr syndrome, Berdon syndrome, Bernard–Soulier syndrome, Beta-ketothiolase deficiency, Beta-mannosidosis, Bietti's crystalline dystrophy, Biotinidase deficiency, Bloom syndrome, Blue diaper syndrome.

CAMFAK syndrome, Canavan disease, CANDLE syndrome, Carbamoyl phosphate synthetase I deficiency, Carey Fineman Ziter syndrome, Carnitine palmitoyltransferase I deficiency, Carnitine palmitoyltransferase II deficiency, Carnitine-acylcarnitine translocase deficiency, Carnosinemia, Carpenter syndrome, Cartilage–hair hypoplasia, Caspase-8 deficiency, Cenani–Lenz syndactylism, Cerebrotendineous xanthomatosis, Chédiak–Higashi syndrome, Chondrodystrophy, Chorea acanthocytosis, Chronic progressive external ophthalmoplegia, Citrullinemia, Cockayne syndrome, Compound heterozygosity, Congenital adrenal hyperplasia, Congenital adrenal hyperplasia due to 21-hydroxylase deficiency, Congenital disorder of glycosylation type IIc, Congenital hepatic fibrosis, Congenital hypofibrinogenemia, Congenital ichthyosiform erythroderma, Congenital insensitivity to pain with anhidrosis, Corneal-cerebellar syndr, me, Cranio–lenticulo–sutural dysplasia, Craniodiaphyseal dysplasia, Cystathioninuria, Cystic fibrosis, Cystinosis, Cystinuria.

De Barsy syndrome, Diastrophic dysplasia, Dicarboxylic aminoaciduria, Dihydropyrimidine dehydrogenase deficiency, Distal spinal muscular atrophy type 1, Donohue syndrome, DOOR syndrome, Dopamine beta hydroxylase deficiency, Dubin–Johnson syndrome, Dubowitz syndrome, EAST syndrome.

EEM syndrome, Ellis–van Creveld syndrome, Endocardial fibroelastosis, Essential fructosuria, Ethylmalonic encephalopathy, Familial dysautonomia, Familial isolated vitamin E deficiency, Familial Mediterranean fever, Fanconi anemia, Farber disease, Fatty-acid metabolism disorder, Fibrochondrogenesis, Finnish heritage disease, Follicle-stimulating hormone insensitivity, Fountain syndrome, Fraser syndrome, Friedreich's ataxia, Fucosidosis, Fumarase deficiency.

Galactokinase deficiency, Galactose epimerase deficiency, Galactose-1-phosphate uridylyltransferase deficiency, Galactosialidosis, Galloway Mowat syndrome, Gangliosidosi, GAPO syndrome, Gastroschisis, Gaucher's disease, Generalized arterial calcification of infancy, Genetic studies on Arabs, Gerodermia osteodysplastica, Giant axonal neuropathy, Gillespie syndrome, Gitelman syndrome, Glanzmann's thrombasthenia, Glucose-galactose malabsorption, Glutaric acidemia type 2, Glutaric aciduria type 1, Glutathione synthetase deficiency, Glycine encephalopathy, Glycogen storage disease type I, Glycogen storage disease type II, Glycogen storage disease type III, Glycogen storage disease type V, Phosphofructokinase deficiency, GM1 gangliosidoses, GM2 gangliosidoses, GM2-gangliosidosis, AB variant, Gonadotropin-releasing hormone insensitivity, Griscelli syndrome, Guanidinoacetate methyltransferase deficiency, Gunther disease.

Harding ataxia, Harlequin-type ichthyosis, Hartnup disease, Heimler syndrome, Hemophagocytic, ymphohistiocytosis, Hereditary folate malabsorption, Hereditary pyropoikilocytosis, Hermansky–Pudlak syndrome, Histidinemia, Holocarboxylase synthetase deficiency, Homocystinuria, Hurler syndrome, Mucopolysaccharidosis type I, Hyperlysinemia, Hypermethioninemia, Hyperprolinemia, Hypertryptophanemia, Hypervalinemia, Hypomagnesemia with secondary hypocalcemia, Imerslund–Gräsbeck syndrome, Iminoglycinuria, Immunodeficiency–centromeric instability–facial anomalies syndrome.

Impossible syndrome, Infantile free sialic acid storage disease, Infantile neuroaxonal dystrophy, Infantile Refsum disease, Infantile systemic hyalinosis, Isobutyryl-coenzyme A dehydrogenase deficiency, Isolated 17,20-lyase deficiency, Isovaleric academia, Jalili syndrome, Jansky–Bielschowsky disease, Jervell and Lange-Nielsen syndrome, Johanson–Blizzard syndrome, Juvenile primary lateral sclerosis.

Kapur–Toriello syndrome, Kaufman oculocerebrofacial syndrome, Keutel syndrome, Kindler syndrome, Kohlschütter-Tönz syndrome, Kostmann syndrome, Krabbe disease, Kufor–Rakeb syndrome, Lafora disease, Lamellar ichthyosis, Laron syndrome, Laurence–Moon syndrome, Lecithin cholesterol acyltransferase deficiency, Lethal congenital contracture syn, Letterer–Siwe disease, Leukocyte adhesion deficiency, Leukocyte adhesion deficiency-1, Leydig cell hypoplasia, Lipoid congenital adrenal hyperplasia, Long-chain 3-hydroxyacyl-coenzyme A dehydrogenase deficiency, Lucey–Driscoll syndrome, Lysinuric protein intolerance, Lysosomal acid lipase deficiency, Lysosomal storage disease.

Malonyl-CoA decarboxylase deficiency, Malpuech facial clefting syndrome, Maple syrup urine disease, Marden–Walker syndrome, Meckel syndrome, Medium-chain acyl-coenzyme A dehydrogenase deficiency, Meleda disease, Metachromatic leukodystrophy, Methemoglobinemia, Methylmalonic academia, Mevalonate kinase deficiency, Michels syndrome, Micro syndrome, Microspherophakia, Microvillous inclusion disease, Mismatch repair cancer syndrome, Mitochondrial DNA depletion syndrome, Mitochondrial neurogastrointestinal encephalopathy syndrome, Mitochondrial trifunctional protein deficiency, MORM syndrome, Morquio syndrome, Mucolipidosis, Mucolipidosis type IV, Mucopolysaccharidosis, Mulibrey nanism, Multiple sulfatase deficiency, N-Acetylglutamate synthase deficiency, Nakajo syndrome, Nemaline myopathy, Nephronophthisis, Netherton syndrome, Neu-Laxova syndrome, Neuronal ceroid lipofuscinosis, Nevo syndrome, Nezelof syndrome, Niemann–Pick disease, Niemann–Pick disease, SMPD1-associated, Niemann–Pick disease, type C, Nijmegen breakage syndrome, North American Indian childhood cirrhosis.

Ochronosis, Oculocutaneous albinism, Oculocutaneous albinism type I, Oculodentodigital dysplasia, Oguchi disease, Omenn syndrome, Opsismodysplasia, Ornithine aminotransferase deficiency, Ornithine translocase deficiency, Orotic aciduria, Otospondylomegaepiphyseal dysplasia, Papillon–Lefèvre syndrome, Pascual-Castroviejo syndrome, Pendred syndrome, Persistent Müllerian duct syndrome, Phenylketonuria, Pipecolic academia, Pontocerebellar hypoplasia, Primary ciliary dyskinesia, Prolidase deficiency, Propionic academia, Pseudo-Hurler polydystrophy, Pseudodominance, Pseudoxanthoma elasticum, Purine nucleoside phosphorylase deficiency, Pycnodysostosis, Pyruvate carboxylase deficiency, Rabson–Mendenhall syndrome, Raine syndrome, RAPADILINO syndrome, Refsum disease, Renal dysplasia-limb defects syndrome, Renal-hepatic-pancreatic dysplasia, Reproductive compensation, Restrictive dermopathy, Roberts syndrome, Rothmund–Thomson syndrome, Rotor syndrome, Sabinas brittle hair syndrome, Salla disease, Sandhoff disease, Sanfilippo syndrome, Sanjad-Sakati syndrome, Sarcosinemia, Sengers syndrome, Senior–Løken syndrome, Short-chain acyl-coenzyme A dehydrogenase deficiency, Shwachman–Diamond syndrome, Sickle cell trait, Sickle cell disease, Situs inversus, Sly syndrome, Smith–Lemli–Opitz syndrome, Spastic ataxia-corneal dystrophy syndrome, Spinal muscular atrophy, Spinal muscular atrophy with progressive myoclonic epilepsy, Spondylo-meta-epiphyseal dysplasia, Spondylo-ocular syndrome, Sporadic late-onset, emaline myopathy, Succinic semialdehyde dehydrogenase deficiency, Sugarman syndrome, Systemic primary carnitine deficiency.

Tangier disease, Tay–Sachs disease, Tetra-Amelia syndrome, Tetrahydrobiopterin deficiency, Thalassemia, Thiamine responsive megaloblastic anemia syndrome, TRIANGLE disease, Trichothiodystrophy, Trimethylaminuria, Triosephosphate isomerase deficiency, Triple-A syndrome, Type I tyrosinemia, Tyrosinemia, Tyrosinemia type II, Tyrosinemia type III, Unverricht–Lundborg disease, Urbach–Wiethe disease, Urocanic aciduria, Urofacial syndrome, Usher syndrome, Very long-chain acyl-coenzyme A dehydrogenase deficiency, Vici syndrome, Walker–Warburg syndrome, Weissenbacher–Zweymüller syndrome, Werner syndrome, Wiedemann–Rautenstrauch syndrome, Wilson's disease, Wolcott–Rallison syndrome, Woodhouse–Sakati syndrome, Xeroderma pigmentosum, Young–Madders syndrome, Yunis–Varon syndrome, Zamzam–Sheriff–Phillips syndrome, ZAP70 deficiency, Zellweger spectrum disorders, Zunich–Kaye syndrome.

X-linked dominant inheritance: Aicardi syndrome, Bazex–Dupré–Christol syndrome, CHILD syndrome, Craniofrontonasal dysplasia, Focal dermal hypoplasia, Fragile X syndrome, Incontinentia pigmenti, Lujan–Fryns syndrome, Oculofaciocardiodental syndrome, Rett syndrome, Template: X-linked disorders, X-linked hypophosphatemia.

X-Linked Recessive conditions: Adrenoleukodystrophy, Becker's muscular dystrophy, Color blindness, Creatine transporter defect, Duchenne muscular dystrophy, Endocardial fibroelastosis, Fabry disease, FG syndrome, Haemophilia, Haemophilia A, Haemophilia B, Hoyeraal–Hreidarsson syndrome, Spinal and bulbar muscular atrophy, Lysosomal storage disease, MASA syndrome, McLeod syndrome, Mendelian traits in humans, Menkes disease, Nasodigitoacoustic syndrome, Norrie disease, Occipital horn syndrome, Ocular albinism, Ocular albinism type 1, Oculocerebrorenal syndrome, Ornithine transcarbamylase deficiency, Pelizaeus–Merzbacher disease, Renpenning's syndrome, Say–Meyer syndrome, Simpson–Golabi–Behmel syndrome, Smith–Fineman–Myers syndrome, Wieacker syndrome, X-linked agammaglobulinemia, Template: X-linked disorders, X-linked dystonia parkinsonism, X-linked intellectual disability, X-linked recessive chondrodysplasia punctate, X-linked spinal muscular atrophy type 2, XMEN disease.

Multifactorial inheritance disorders: There are many genetic conditions that are controlled by multiple genes mostly but not always in interaction with environment. Examples are: Congenital malformation, Cleft palate, Congenital dislocation of the hip, Congenital heart defects, Neural tube defects, Pyloric stenosis, Talipes, Adult onset diseases, Diabetes Mellitus, Cancer, Glaucoma, Hypertension, Ischaemic heart disease, Manic depression, Schizophrenia, Psoriasis, Thyroid diseases, Alzheimer's Disease. Expression of some of these diseases can be changed or controlled by diet and specific herbals.

Mitochondrial Genetic Disorders

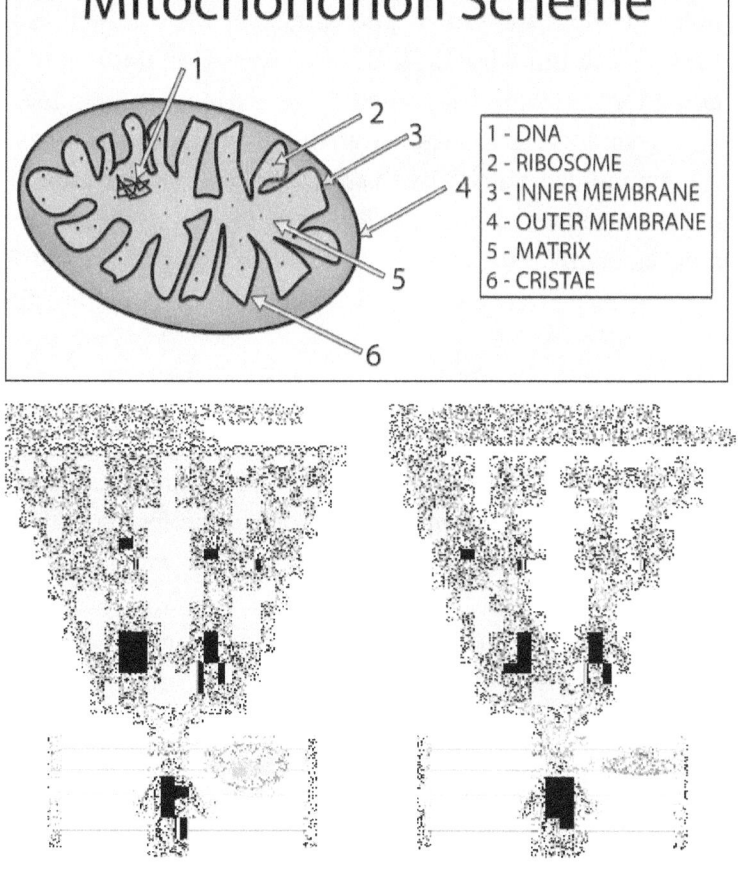

Mitochondria are small bacteria like structures found in every human and other eukaryotic cell except red blood cells. There are about 100 mitochondria in each human cell. Each mitochondrion has an average of 210 copies of circular double stranded DNA having 16569 base pairs of the DNA. Mitochondria are essential for sustaining life since they are the powerhouses wherein energy for conducting all bodily functions in the form off ATP molecules is synthesized. Mitochondrial DNA is not transmitted in a Mendelian fashion unlike nuclear DNA. Mitochondrial DNA is inherited only from the mother's ovum. 80% off mitochondrial DNA codes for functional mitochondrial proteins. The proper functioning of mitochondria required both normal mitochondrial DNA coded proteins as well as nuclear coded proteins. In fact, only 13 of the 3000 proteins found in the mitochondrion are coded by the mitochondrial DNA the rest being coded by nuclear DNA. Thus, defects in the mitochondria can be due to both mutations in mitochondrial DNA as well as nuclear DNA. Mutations in mitochondrial DNA as well as mutations in nuclear genes coding for mitochondrial functions lead to myopathies and muscle disorders. These are referred to as mitochondrial neuropathies as well as neurological manifestations. The reason for such manifestation appears to be due to muscles and neurological cells requiring more energy for functioning as compared to other cells. Since muscles are, an integral part of our organs, mutations in mitochondrial DNA can affect all bodily functions.

Mitochondrial disease is a group of disorders caused by dysfunctional mitochondria and primarily affects the brain, nerves, muscle, and heart and all those organs that require proper muscular function. Because muscular function is a requirement for proper functioning of all organs, symptoms due to mitochondrial dysfunction may mimic diseases like autism, Parkinson's disease, Lou Gehrig's disease, Muscular dystrophy, Alzheimer's disease, visual and hearing problems, gastrointestinal disorders, heart functioning etc. all of which may have other causes. In general, symptoms caused by a mitochondrial genetic disease may include growth abnormalities, lack of muscle coordination, weak muscles, learning disabilities, mental retardation, heart and kidney disease, gastrointestinal disorders, respiratory problems, and diabetes.

Some of the more common mitochondrial myopathies include Kearns-Sayre syndrome, myoclonus epilepsy with ragged red fibers and mitochondrial encephalomyopathy with lactic acidosis and stroke like episodes. The symptoms of mitochondrial myopathies include muscle weakness or exercise intolerance, heart failure or rhythm disturbances, dementia, movement disorders, stroke like episodes, deafness, blindness, droopy eyelids, limited mobility of the eyes, vomiting and seizures. The prognosis for these disorders ranges in severity from progressive weakness to death. Most mitochondrial myopathies occur before the age of 20 and often begin with exercise intolerance or muscle weakness. During physical activity, muscles may become easily fatigued or weak. Many diseases of aging are due to mitochondrial dysfunction because mitochondria handle using oxygen and converting glucose into energy.

Mitochondrial DNA is susceptible to damage from free radicals that can damage mtDNA. Such mutations can accumulate with aging particularly in oxygen dependent tissues such as brain, heart muscle, and kidneys. The following is a list of major diseases that have been found as having involvements of mitochondrial functional genes. All have muscle and or neural involvement due to impairment of energy generation. Examples are: KearnsSayre syndrome, Leber hereditary optic neuropathy, Leigh syndrome, Maternally inherited diabetes and deafness, Mitochondrial encephalomyopathy, lactic acidosis and stroke like episodes, Mitochondrial complex III deficiency, myoclonic epilepsy with ragged red fibers, Non syndromic deafness, Neuropathy ataxia and retinitis pigmentosa, Pearson marrow pancreas syndrome and progressive external ophthalmoplegia.

There is no specific treatment for controlling diseases caused by mitochondrial defects. However, since oxygen free radicals may be playing a role in expression of mitochondrial defects especially during the aging process it may be possible to reduce the impact of age based expression of disease conditions by consuming foods that may prevent formation and accumulation of free radicals in the body. Many vegetables, fruits and other plant-based products are useful in scavenging free radicals from the body and hence reducing the possibility of the aging process itself.

References

https://www.genome.gov/10001204/specific-genetic-disorders.

https://www.genome.gov/27527652/genomic-medicine-and-health-care.

https://www.rxlist.com/genetic_disease/article.htm.

https://www.healthdirect.gov.au/genetic-disorders.

https://www.stanfordchildrens.org/en/topic/default?id=types-of-genetic-diseases-90-P02505.

https://www.britannica.com/science/human-genetic-disease/Management-of-genetic-disease.

Cummings, M.R. (1988): Human Heredity, Principles and Issues, West Publishing Company, St Paul, Minn, USA, pp 475.

Lewis, R. (2017): Human Genetics, 12th edition, Mc Graw Hill publishing.

William S Klug, Michael R Cummings, Charlotte A Spencer, Michael A Palladino, Darrell Killian (2019) Concepts of Genetics, 12th edition, Pearson publishing.

CHAPTER 5

Cancer, Herbs, Chemoprevention

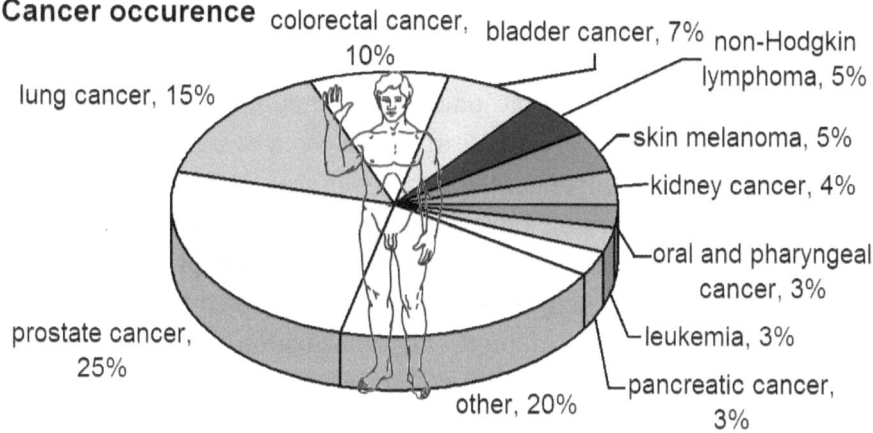

Cancer is a very general term covering over 100 types of diseases resulting from uncontrolled cell division, growth and spread of these abnormal cells to different parts of the body. The most common cancers are, Bladder, Breast, colon, Endometrial, Kidney, Leukemia, Liver, Lung, ovarian, Non-Hodgkin Lymphoma, Skin, Pancreatic, Prostate and Thyroid cancers.

Table 7: lists the different types of cancers symptoms and genes involved. While some of these cancers result in tumorous growth, others like leukemia cause proliferation of leucocytes (white blood cells) without tumor formation. Certain types of uncontrolled cell division result only in localized tumors and such tumors are said to be benign. Yet others are malignant and spread from their point of origin to other parts of the body by metastasis

When the human embryo develops, into the adult it undergoes cell division and organogenesis in a precisely controlled fashion so that the proper genetically determined morphology of the individual is kept. Some of the genes involved in this precise pattern formation and development are called **proto-oncogenes** and yet others are **tumor suppressor** genes. All humans and animals have proto-oncogenes and tumor suppressor genes that are highly active in cell division, pattern formation, organ formation and cell death -apoptosis. However, after the formation of the adult body, many of these proto-oncogenes are shut off while a few others are still active. Examples of proto-oncogenes include receptor tyrosine kinases eg: Brawl membrane associated proteins, transcription factors, Myc, and different growth factors. When these proto-oncogenes undergo mutation due to point mutations, deletion or insertion mutations, amplification of genes, reactivation of silent proto-oncogenes or inhibition of cell apoptosis, such cells are destroyed. Nevertheless, if such changes are ignored and the cells carrying these changes continue to divide then, the proto-oncogenes become the cancer-causing oncogenes. In addition, tumor suppressor genes that in normal cells prevent the transformation of proto-oncogenes into oncogenes become marginalized due to mutations then, these tumor suppressor genes are unable to prevent transformation of proto-oncogenes into oncogenes. Further, reactivation of proto-oncogenes that were turned off earlier in life could also contribute to cell transformation from normal to abnormal. Most normal cells undergo a programmed form of rapid cell death known as apoptosis. Activated oncogenes can cause those cells chosen for apoptosis to survive and proliferate instead causing malignancy.

Such transformation of proto-oncogenes into oncogenes can occur due to certain inherited factors or due to chemicals or radiation and biological infectious agents like viruses. Thus, point mutations, translocations, gene amplifications, gene deletions and insertions can cause transformation of cells into abnormality. The expression of oncogenes is regulated by small RNA molecules known as miRNA or Oncomirs and mutations in these RNAs can also lead to expression of oncogenes. When such oncogenes are generated, abnormal proteins are produced and cells that should normally die continue to divide and spread from their place of origin to other sites resulting in malignancy. Thus, transformation of cells towards cancer is an overly complex process that is tightly regulated and when this regulated cell activity is disturbed then cells become malignant.

Many different cancers are triggered by infections particularly by viruses such as adeno, herpes, Hepatitis B, Hepatitis C, polyoma, papilloma, human T cell lymphotropic, and other retroviruses. These virus induced cancers are preventable through immunization wherever vaccines are available.

Several plants have been examined for their potential as anticancer agents. Many programs funded by Governmental and private agencies take part in finding potential anti-cancer phytochemicals and plants having these phytochemicals. The most ambitious such program was started by the National Cancer Institute of the United States of America. Initially, the program used a random method for identifying antitumor activity wherein, the single major constituent of the selected plants were chosen without taking into full consideration that plants contain numerous phytochemicals in a single plant and that very often it is the collective action of all components in a plant that were biologically active. Most ancient forms of medicine such as traditional Chinese medicine and the Indian Ayurvedic medicine adopted a holistic method whereby the whole plant or plant part rather than any single part of the plant was prescribed as a treatment. The traditional practitioners including herbalists considered the fact that the total therapeutic activity of a plant extract is greater than or different from the therapeutic activities of the individual chemicals. It is common for a fraction from a plant extract that has significant biological activity to have many constituents with this activity and act in a synergistic fashion.

Plants such as the periwinkle do have active chemicals that have been extracted, evaluated and are now in treatment regimens. Yet other plant-based products are being evaluated for *invitro* and *in vivo* activities. Evidence is also accumulating that the incidence and severity of cancer can be reduced by exercise, meditation, and proper nutrition. The role of plant Nutraceuticals in cancer treatment and or prevention is not fully evaluated. Fruits, vegetables, and grains rich in antioxidants, fiber and other phytochemicals seem to help in cancer prevention. In this Chapter, a brief description of plants having phytochemicals with proven anticancer and potential activity as well as those considered to have anticancer activity by traditional medical practice will be given.

Podophyllotoxin   Taxol

Campothecin   Vinblastin   Vincristin

Madagascar periwinkle/Nithya Kalyani/Sadabahar (*Catharanthus roseus*): Kingdom: Plantae, Group: angiospermae, Eudicots, Order: Gentianales, Family: Apocyanaceae

The most important plant source of materials used in chemotherapy is the alkaloids of Madagascar periwinkle *Catharanthus roseus*. This plant was used traditionally for treatment of diabetes but later the Vinca alkaloids were found to have anti-leukemic properties. Vincaleukoblastine, vinblastine, vincristine, and leurocristine from this plant are now used either alone or in combination with other forms of therapy for cancer treatment. Vinblastine is useful in the treatment of Hodgkin's disease a cancer affecting the lymph glands, spleen, and liver whilst vincristine is used for childhood leukemia. Vinblastine, Vincristine, Vinorelbine and vindesine are listed among the essential drugs by the WHO.

Periwinkle plants are evergreen herbs grown mostly in tropical and subtropical climates but grown as an annual in temperate countries. Leaves are about 1 inch to 5 inches long, glabrous arranged in opposite pairs. Flowers are white to pink in nature but hybrids with other colors are available by commercial breeding.

Mayapple/American mandrake/wild mandrake/and ground lemon (*Podophyllum peltatum, P. hexandrum*): Kingdom: Plantae Group: Angiospermae, Eudicots. Order: Ranunculales, Family: Berberidaceae

Podophyllum are woodland plants typically growing in colonies derived from a single root. The plants grow to about 12 ft high with palmate lobed umbrella-like leaves up to 6 inches-8 inches in diameter. The plants produce several stems from a creeping underground rhizome. Some stems bear a single leaf and do not produce any flower or fruit while flowering stems produce a pair or more leaves with 1–8 flowers in the axil between the apical leaves. The flowers are white yellow or red.

Health relevance: *Podophyllum* has lignans known as Podophyllotoxins, which have antitumor activity. The major natural constituent Epipodophyllotoxin is unsuitable for use due to high toxicity. Nevertheless, two derivatives namely Etoposide and Teniposide have given good results in clinical trials. Etoposide is available for the treatment of small cell lung cancer and testicular cancer and Teniposide is used in pediatric cancers. Apart from anti-cancer activities, Podophyllum rhizomes are used in American Indian folk medicine as an emetic and anti-helminthic.

***Pacific yew (Taxus brevifolia,* Canadian yew (*Taxus canadiensis*), European Yew (*T. baccata*), Kingdom: Plantae, Clade: Gymnospermae, Pinopsida, Order: Pinales, Family: Pinaceae.**

The plants belong to the Yew family. They are slow growing and can be very long-lived and reach heights of over 100 ft. with trunk diameters of

up to 15-20 ft. They have reddish bark and lanceolate flat dark green leaves arranged spirally on the stem. The bark of the Pacific yew was the original source of anti-cancer Taxol but it is now known that leaves of the European yew also contains the Taxol's which are the basis of the semi-synthetic Taxol group of anti-cancer drugs.

Health Relevance: **Taxane/Paclitaxel/Taxol/Docetaxel.** Paclitaxel and docetaxel are the two presently clinically available representatives of the new class of taxane drugs. They share major parts of their structures and mechanisms of action but differ in several other aspects. For instance, there is a difference in their tubulin polymer generation and docetaxel appears twice as active in de-polymerization inhibition. Docetaxel that is marketed under generic or under the trade names Taxotere or Docecad is a clinically well-established antimitotic chemotherapy medication that works by interfering with cell division. Docetaxel is approved by the FDA for treatment of locally advanced or metastatic breast cancer, head and neck cancer, gastric cancer, hormone refractory prostate cancer and non-small-cell lung cancer. Docetaxel can be used as a single agent or in combination with other chemotherapeutic drugs as shown, depending on specific cancer type and stage. Paclitaxel is a medication used to treat several types of cancer including ovarian cancer, breast cancer, lung cancer and pancreatic cancer, among others. It and docetaxel represent the taxane family of drugs. Paclitaxel's mechanism of action involves interference with the normal breakdown of microtubules during cell division. Paclitaxel is currently listed among the essential drugs by the WHO.

Taxane are also used in preparing drug-eluting stents that are placed by invasive cardiologists and radiologists in arteries to prevent cell proliferation thus preventing clotting and blockage of arteries.

Happy tree/tree of Life Xi Shu *(Campotheca acuminata):* **Kingdom: Plantae, Clade: Angiospermae, Eudicots, Order: Cornales, Family: Cornaceae.**

Camptotheca is a genus of medium sized deciduous trees native to southern China and Tibet. They grow to about 70 ft. tall. Leaves are broad and resemble avocado leaves. The bark leaves and seeds are used for medicinal purposes particularly for treating cancers. The bark is grey in color and smooth when young but becomes furrowed as the tree ages. Blooms are white or pale green, yellow in color, spherical in shape resembling a puffy marble and are borne in clusters in the spring. Blooms are followed by fruits with wings also borne in spherical clusters. Leaves are ovate in shape and are arranged opposite one another.

Health relevance: *The* following drugs are derived from this plant. They are respectively, Campothesin, Topotecan, Hycamtin, Irinotecan, Camptosar, Campto, and CPT11. They are all approved drugs available by prescription for cancer treatments.

Cowtail Pine/Japanese Plum/Yew *(Cephalotaxus harringtonia):* **Kingdom: Plantae Clade: Gymnospermae, Pinopsida, Order: Pinales, Family: Cephalotaxaceae.**

Plants are coniferous evergreen trees native to Japan but grown in Europe and elsewhere mostly as ornamental trees. Plants are dioecious separate male and female plants. Leaves are linear and fan shaped.

Health relevance: Harringtonine and homoharringtonine are naturally occurring alkaloids found in the Japanese plum yew with proven antineoplastic activity against certain types of leukemias in cell cultures, experimental animals, and initial clinical trials. The drug SYNRIBO (Omacetaxine mepesuccinate) which is a derivative of homoharringtonine has been approved for treatment of chronic mylogenous leukemia.

Yellow wood (*Bleekeria vitensis/Ochrosia vitensis/O. moorei and Excavatia coccinea*) Kingdom: Kingdom: Plantae, Clade: Angiospermae, Eudicots, Order: Gentianales, Family: Apocyanaceae. Plants are herbs/shrubs native to South Asia, Indian Ocean islands, Australia, New Guinea, and Indonesian archipelago. **Health Relevance:** Elliptinium/ Ellipticine found in Ochrosia exhibit significant antitumor and antihuman immunodeficiency virus activities. The alkaloids inhibit DNA replication and protein synthesis.

Indian White Cedar/Bombay white cedar (*Disoxylum binectariferum*): Kingdom: Plantae, Clade: Angiospermae, Eudicots, Sapindales, And Meliaceae

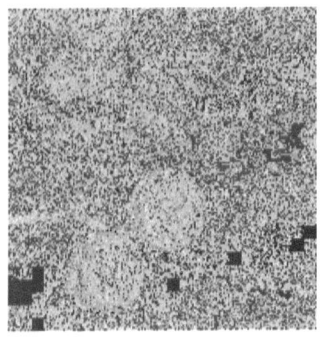

These are trees growing in India, China, New Caledonia, and many Asian rain forests. The plant has phytochemical flavanoids with potential use as anti-cancer agents. Rohitukine found in this plant is an important precursor for the semi-synthetic derivative, flavopiridol. Flavopiridol (Alvocidib) is a semi-synthetic flavonoid based on an extract from *Disoxylum* for the potential treatment of cancer. It works by inhibiting cyclin-dependent kinases, arresting cell division and causing apoptosis in non-small lung cancer cells. It has been studied for the treatment of acute myeloid leukemia, arthritis, and atherosclerotic plaque formation. It has a role as an antineoplastic agent. Modified forms of Flavopiridol known as Voruciclib and Dinaciclib are claimed to show a more selective target profile than flavopiridol. Flavopiridol has been given Orphan drug approval by the FDA. (The Orphan Drug Act (ODA) provides for granting special status to a drug or biological product ("drug") to treat a rare disease or condition upon request of a sponsor. This status is referred to as orphan designation (or sometimes "orphan status").

Narkia tree/Kalagaura (*Nathopodites foetida/Mappia foetida*): Kingdom; Plantae, Clade: Angiospermae, Eudicots, Order: Icacinales, Family: Icacinaceae

Mappia foetida locally known as Narkia or Amrita is an endangered plant species naturally distributed in the Western Ghats and Maharashtra state of India. Other related species are also grown in China and is a part of Chinese TCM. It is the most convenient source for large-scale production of the anti-cancer drug Camptothecin (CPT). The other source is the Cancer tree *Campotheca* as described above. Two CPT analogues topotecan and irinotecan have been approved for cancer chemotherapy.

Pau d'arco/Lapacho/Trumpet tree (*Tabuia avellanedae/Handroanthus impetiginosus*). Kingdom; Plantae, Clade: Angiospermae, Eudicots, Order: Lamiales, Family: Bignoniaceaee,

This plant is a deciduous native tree of South America, where the bark and stem of the trees has been used to treat a wide range of conditions including cancer, arthritis, pain, inflammation of the prostate gland, fever, dysentery, boils, ulcers, and fungal infections. The inner bark of these trees is used to make a tea infusion much like a tonic for treatment. There are reports of medicinal uses of pau d'arco tea dating back to 1873.

Pau d'arco/Lapacho is a tall tree with thick trunks and trumpet-like flowers. The inner bark (phloem) has the phytochemical Lapachol and beta-lapachone. Pau d'arco tea has several compounds, including quinoids, benzenoids and flavonoids. These compounds have shown biological activity against harmful organisms. A wide range of pharmacological activities has been seen for Lapachol and its semi-synthetic derivatives. In the case of cancer, some studies showed that it is effective in reducing metastasis of cancer cells. However, recent studies show that the drug is very toxic and hence its use for consideration as an anti-cancer agent has been stopped. The extracts are Abortifacients. According to the American Cancer Society, "available evidence from well-designed, controlled studies does not support this substance as an effective treatment for cancer in humans".

Plants with Potential Anti-Cancer Chemicals

The world-wide literature on plants and phytochemicals that have anti-cancer properties is large, in fact about every plant has some anti-cell proliferative, or anti-tumor property when evaluated in vitro or in animal models. The description below is the result of sifting through reviews and other literature and is not a complete list of plants and phytochemicals with anti-cancer properties.

Barberry (*Berberis vulgaris, B. aristata, B. hydrastis*): Kingdom: Plantae, Clade: Angiospermae, Eudicots, Order: Ranunculales, Family: Berberidaceae

This plant has alkaloid **Berbamine** isoquinolone. This alkaloid shows Caspase 3 inhibition. It modulates angiogenesis (development of new blood vessals) and modulates tumor necrosis factor TNF (cell death). It is part of Chinese Herbal medicine and is used by traditional Medical practitioners to treat chronic Myeloid Leukemia, breast, liver, bone, prostate, pancreatic, neuroendocrine, Skin, oral and other cancers.

Bittersweet Nightshade/Bitter nightshade (*Solanum dulcamara*): Kingdom: Plantae, Clade: Angiospermae, Eudicots, Order: Solanales, Family: Solanaceae

This plant has **beta solamarine** that shows tumor-inhibitory activity against Sarcoma 180 in mice. Chinese ancestors have used extracts from the Solanum species, including *Solanum lyratum*, and *Solanum nigrum* L. to treat human cancers before 600 A.D. In western world, *Solanum dulcamara* L. has been widely used in folk medicine to treat cancers and warts from the time of Galen (circa A.D. 180). The anti-cancer properties of the extracts from the Solanum species have been known for decades.

Black caraway/Black cumin/Karum Jeerakam (Tamil) (*Nigella sativa*) Kingdom: Plantae, Clade: Angiospermae, Eudicots, Order: Lamiales, Family: Ranunculaceae

Nigella is an annual herbaceous plant native to South West Asia and cultivated and naturalized in Europe and North Africa. Nigella seeds have been used as a spice and a condiment. In traditional medicine, the seeds have been used in different forms to treat many diseases including asthma, hypertension, diabetes, inflammation, cough, bronchitis, headache, eczema, fever, dizziness, and influenza.

Recent research reports have shown that N. *sativa* is a dietary supplement (DS) in complementary and alternative medicine (CAM) along with chemotherapy. The major constituent **Thymoquinone** is a bioactive compound, which has been shown to have antineoplastic activity against multifarious tumors including lung, breast, pancreatic, colorectal, and other cancers.

Cancer weed/milkweed (*Euphorbia peplus*): Kingdom: Plantae, Clade: Angiospermae, Eudicots, Order: Euphorbiales, Family: Euphorbiaceae.

This plant exudes a sap having a terpene known as **Ingenol mebutate** that is used as a constituent in skin creams for treatment of actinic keratosis. Actinic keratosis also called solar keratosis is a crusty, scaly growth caused by damage from exposure to ultraviolet (UV) radiation. It is a pre-cancerous skin condition. Ingenol mebutate is an FDA approved gel for topical applications and sold under the proprietary name Picato. The plant sap has been used by traditional herbalists as a treatment for skin lesions.

This cancer weed is an invasive species distributed all over the world but originally this species is a native of Europe and Mediterranean countries where it grows in wasteland. Plants are annuals about 1 ft- 5 ft tall with ovate leaves and umbellate green flowers.

Colchicum (*Colchicum autumnale*): Kingdom: Plantae, Clade: Angiospermae, Monocots, Order: Liliales, Family: Colchicaceae. The alkaloid **Colchicine** affects Mitotic replication and angiogenesis and has anti-neoplastic activity. The allopathic drug **colchicine** from this plant is prescribed for controlling gout.

Ginger (*Zingiber officianalis*): Kingdom: Plantae, Clade: Angiospermae, monocots, Order: Zingiberales Family: Zingiberaceae

Extracts of ginger rhizome causes induction of apoptosis of human prostate cancer cells *in vitro*. Growth of Tumor tissue from mice was suppressed and it did not show any detectable toxicity in normal tissues. One of the main ingredients is **Gingerol**.

Gourds/Cucumbers (*Cucurbita* sp): Kingdom: Plantae, Clade: Angiospermae, Eudicots, Order: Cucurbitales, Family: Cucurbitaceae.

The phytochemical **Cucurbitacin** in cucurbits has antioxidant, antiinflammatory, and anticancer activity in cell lines. All members of the Cucurbitaceae family are annual vines or climbers with tendrils but a limited few are shrubs or short trees. In general, they produce unisexual flowers on the same plant or on separate male and female plants. They produce typical fruits known as gourds or squashes or pumpkins.

Miracle Berry/Miracle fruit (*Synsepalum dulcificum*): Kingdom: Plantae, Clade: Angiospermae, Eudicots, Order Ericales, Family: Sapotaceae.

This red colored oval fruit of this plant produces a glycoprotein known as **Miraculin**. Although not a sweetener by itself, Miraculin binds to the taste receptors in the tongue/mouth and makes tart and metallic liquids taste sweet. If patients chew the miracle berry or its extracts before medicating then, it overcomes the metallic taste of oral chemotherapy agents. The plant from which this glycoprotein is extracted is a native of Africa. It is a shrub with glabrous leaves. It is grown commercially in Florida, USA. The Miraculin gene has been cloned and expressed in E. Coli and in Lettuce.

South African Bush Willow: (*Combretum caffrum*): Kingdom: Plantae, Clade: Angiospermae, Eudicots, Order Myrtales, Family: Combretaceae.

This plant, having the chemical **Combretastatin** with anti-tubulin activity in experimental cell line, is a candidate for developing anticancer drugs. The drugs, Combretastatin A1 and A4 - based on a tree bark extract cuts the supply of oxygen-laden blood to the cancers without causing serious side effects to patient.

The following plants that have also been evaluated for anti-cancer activity have shown some level of anti-tumor cell activities. They are *Actaea racemosa, Ardisia crenata, Bacopa Monnieri, Cedrus deodara, Cyanodon dactylon, Piper longum, Picrorhiza kurroa, Plumbago zeylanica, Punica granatum, Thymus vulgaris, Tinospora cordifolia* and *Zanthoxylum nitidum*.

Table 7: List of plants from which approved anticancer chemicals have been used to develop anticancer therapeutic drugs

Common Names	Botanical Name	Active Chemicals	Semisynthetic	Chemical class	Mode of action
Periwinkle	*Catharanthus roseus Madagascar*	Vincristine, Vinblastine	Vinorelbine vindesine	Alkaloids	Antimitotic
American May Apple	*Podophyllum peltatum*; *P. Emodi*; *P. Hexandrum*	Podophyllotoxin	Etoposide, Teniposide	Lignans	DNA breakage by inhibition of Topoisomerase II
Pacific Yew; Common Yew	*Taxus brevifolia T. baccata*	Plaxitaxel Taxol	Docetaxel	Taxanes	Mitosis inhibitor
Happy TreeXi Shu	*Camptotheca acuminata*	Camptothecin	TopotecanIrenotecan, Exatecan, Lurtotecan, CZ48, Rubitecan, CPT 11	Quinoline alkaloids	DNA replication inhibits Topoisomerase
Nothopodytes Tree Narkia Tree	*Nothapodytes Foetida*	Camptothesin Acetyl Campothesin Scopolectin		Quinoline alkaloids	Inhibits DNA Topoisomease

Anti-Cancer Phytochemicals

The following is a list and brief description of phytochemicals with potential anticancer properties based on *invitro* and animal studies. All the studies have been conducted *in vitro* in human and or animal cell cultures. The data from these studies could be the basis for developing delivery systems and formulations.

Andrographolide ($C_{20}H_{30}O_5$): (*Andrographis paniculata*): Extracts of the plant having **Andrographolide** exhibited cytotoxicity against a panel of cancer cell lines. Andrographolide regress tumor in mice.

Apigenin ($C_{15}H_{10}O_5$): Parsley (*Petroselinum crispum*), Celery (*Apium graveolens*), Chamomile (*Matricaria recutita and Chamaemelum nobile*) and Moringa (*Moringa* sp.). Apigenin is a flavone present in vegetables such as parsley, celery, chamomile, and Egyptian plant *Moringa peregrin*. Apigenin exhibits apoptic properties against breast cancer cell lines and colon cell lines comparable to that of doxorubicin. It is suggested as a chemoprevention agent.

Barberry (*Berberis sp*): Has **Berbamine** alkaloid ($C_{37}H_{40}N_2O_6$). This chemical shows Caspase 3 inhibition. It is used in Chinese herbal for chronic myeloid leukemia.

Betulinic acid ($C_{30}H_{48}O_3$): White birch (*Betula Alba, B. pubescens*), Jujube (*Ziziphus nummularia,*). Betulinic acid initiates apoptosis of melanoma cell lines, and takes part in Caspase activation, mitochondrial membrane disruption and DNA breaks. Betulinic acid is not soluble in aqueous solutions.

Lepacho tree (*Tabebuia avellanedae*): This tree produces **Beta-lapachone** ($C_{15}H_{14}O_3$) that shows DNA topoisomerase inhibitor activity. It has great potential for the treatment of specific cancers with elevated NQO1 levels (e.g., breast, non-small cell lung, pancreatic, colon, and prostate cancers). This phytochemical is extremely toxic.

Capsaicin ($C_{18}H_{27}NO_3$): Red Chili pepper (*Capsicum* sp.). Capsaicin controls pain and is a part of many pain-killing topical medications. However, many studies suggest that it has anti-tumor properties. *In vitro* cell culture studies showed that capsaicin induced inhibition of cell proliferation of human gastric cell lines. It also targeted angiogenesis in non- small cell lung carcinoma cells. It also promoted apoptosis of small cell lung carcinoma cells. However, there are many other reports based on epidemiology, which correlate the consumption of red chillies with gastrointestinal cancers.

Turmeric (*Curcuma longa*): Kingdom: Plantae, Clade: Angiospermae, Monocots, and Order: Zingiberales, Family: Zingiberaceae. The plant biology details are given in the chapter on digestive diseases. Curcumin and turmerone oil, Curcumindemethoxycurcumin, bisdemethoxycurcumin, Alphaturmerone and turmerone are the main constituents of turmeric. The extracts of the rhizome are extensively used in Ayurvedic traditional medicine to enhance immune activity and is considered immune protective from colon cancers and polyps.

Turmeric shows several therapeutic effects, which can play a role in treating Nasopharyngeal, Lung, hepatobiliary, breast, gastric and colorectal cancers. Thus, experiments have shown that curcumin arrests G2 cell cycle, kill human nasopharyngeal cell lines, and increase the radio sensitivity of these cells. In addition, some studies in animals have shown the anti-tumor effects of curcumin in lung cancer. Other studies have shown that curcumin increases the apoptotic rate in breast hepato biliary and gastric cancers. Similarly, application of curcumin-containing vaginal creams on the vaginal epithelium eradicated Herpes virus (HPV) from HPV+ cells. **Note:** Curcumin is poorly absorbed by the cells and as such, it is said that huge quantities of turmeric must be prescribed to achieve any metabolic effect. However, addition of **piperine** from black pepper or preparation of liposomal curcumin can increase absorption. Many recent investigators including the NIH report that the medicinal property of curcumin has been exaggerated and that clinical evidence of anti-cancer or anti-inflammatory activity is lacking. More studies are needed to verify the anti-cancer effects of curcumin before it can become a regular treatment regimen.

Crocin ($C_{44}H_{64}O_{24}$) and crocetin ($C_{20}H_{24}O_4$): Saffron flower stamens (*Crocus sativus*), Family: Iridaceae. More than 60 journal publications dwell on the therapeutic and chemo-protective nature of **crocin** and crocetin. Several studies showed that saffron had anticancer effects that were attributed to the bioactive compounds it had. Saffron and its carotenoids exert chemo-preventive activity through antioxidant activity.

Cyanidin ($C_{15}H_{11}O_6+$): Grapes (*Vitis vinifera*), blackberry (*Rubus fruticosus*), cranberry (*Vaccinium erythrocarpum, V. macrocarpon, V. microcarpum, V. oxycoccus*) raspberry (*Rubus idaeus*), apples (*Pyrus malus*, plums (*Prunus*), red cabbage (*Brassica oleracea* Capitata Group) and red onion (*Allium cepa*)

Cyanidin has antioxidant activity that could help in cancer prevention as shown in colon cancer and breast cancer cell-lines as well as rat esophageal cell lines.

Diindolylmethane (DIM, $C_{17}H_{14}N_2$) and Indole-3-carbinol (I3C, C_9H_9NO) from Brassica vegetables (Broccoli, collard greens, Cauliflower). This indole compound inhibits hormone responsive cancer cells of breast, prostate and ovarian cancers as well as tobacco induced adeno carcinomas of the lung. **Indole-3-carbinol (I3C)** is found in Brassica vegetables, such as broccoli, cauliflower, collard greens. Diindolylmethane (DIM) is a digestion derivative of indole-3-carbinol via condensation formed in the acidic environment of the stomach. I3C has been studied for cancer prevention and therapy for years. I3C and DIM proved exceptional anti-cancer effects against hormone responsive cancers like breast, prostate and ovarian cancers.

Ellipticine ($C_{18}H_{16}N_2O$): Apocynaceae (*Ochrosia sp.*): Ellipticine is an intercalator of DNA, inhibits DNA topoisomerase II, exhibits *invitro* activity against MCF7 cells, leukemia HL60 and CCRF-CEM cells, neuroblastoma IMR32 UKFNB3 and UKFNB4 cells and U87MG glioblastoma cells

Emodin ($C_{15}H_{10}O_5$): Rhubarb (*Rheum emodi, R. Palmatum*): Emodin initiates apoptosis in human cervical cancer HeLa cells.

Epifriedelinol (Vitex peduncularis): Verbenaceae: Extracts of Vitex have many chemicals of which **Epifriedelinol** has been studied and found to induce apoptosis of cervical cancer cells.

Epigallocatechin ($C_{15}H_{14}O_7$): Tea (*Thea sinensis*): This phytochemical from tea especially green tea inhibits and kills cell lines from brain, prostate, cervical, colon and bladder cancers.

Fisetin ($C_{15}H_{10}O_6$): Apples (*Pyrus malus*), Strawberries (*Fragaria sp.*) persimmon (*Diospyros, sp.*) grape (*Vitis vinifera*), onion (*Allium cepa*), cucumber (*Cucumis sativus*), Acacia (*Acacia regia, A. berlandieri,*): This flavanoid exerts anti-carcinogenesis effects in HCT-116 human colon cancer cells as well as antiinflammatory effects in many cell lines.

Genestein ($C_{15}H_{10}O_5$), Soy beans (*Glycine max*), Lupine (*Lupinus sp.*), fava beans (*Vicia faba*), coffee (*Coffea arabica*): This isoflavone originates from a number of plants such as lupine, fava beans, soybeans, kudzu, psoralea, and coffee. Functioning as antioxidant and anthelmintic, Genistein has been found to have anti-angiogenic effects -blocking formation of new blood vessels and choking of blood supply to cancer cells.

Galangin flavonol ($C_{15}H_{10}O_5$): Lesser galangal (*Alpinia officinarum*), Zingiberaceae: This flavone isolated from the rhizomes of the galangal plant has been shown to have anti-tumor activities including melanoma, hepatoma, and colon cancer cells.

Gingerol ($C_{17}H_{26}O_4$) (Ginger, *Zingiber officianalis*): Gingerol has proven antioxidant, anti-inflammation, and antitumor promoting properties in animal and invitro studies.

Kaempferol ($C_{15}H_{10}O_6$): Broccoli (*Brassica oleraceae*), Witch Hazel (*Hamamelis sp.*), grapefruit (*Citrus × paradisi*), Brussel sprouts (*Brassica oleracea*): Kaempferol is a natural flavonol present in above plants. It has been investigated for its anti-angiogenic, anticancer, and radical scavenging effect on pancreatic cancer, and lung cancer cell lines in vitro. **Kaempferol,** displayed moderate cytostatic activity in the cell lines of PC3, HeLa and K562 human cancer cells.

Lycopene (C40H56): Tomato (*Lycopersicon esculentum*), carrots (*Daucus carota*), Watermelon (*Citrullus lanatus*): Lycopene is a bright red pigment, which has antioxidant activity and Chemopreventive effects on prostate cancer cells in animal studies. The anti-cancer property is attributed to activating cancer preventive enzymes. Lycopene was found to inhibit human cancer cell proliferation, and to suppress insulin-like growth factor-I-stimulated growth. This suggests a role for lycopene in the prevention or treatment of endometrial cancer, prostate and colon cancer.

Moronic acid: Sumac (*Rhus Javanica*), Mistletoe ((*Phoradendron reichenbachianum*): Moronic acid is a triterpene that is cytotoxic with potential anti-cancer activity, it is active against HIV and Herpes simplex viruses.

Phenethyl Isothiocyanate (PEITC, C9H9NS): Cruciferous vegetables

PEITC, along with sulforaphane from cruciferous vegetables, such as watercress, Broccoli, cabbage, etc., have been studied for induction of apoptosis in cell lines. PEITC has shown strong potency against melanoma. It has been intensively studied for chemoprevention against breast cancer cells, non-small cell lung cancer, cervical cancer, osteogenic sarcoma, and prostate cancer and myeloma cell lines. PEITC induces apoptosis in some cell lines that are resistant to some currently used chemotherapeutics drugs.

Piperine (C17H19NO3) Black pepper (*Piper nigrum*):

Piperine is the major bioactive compound in black pepper. Black pepper is one of the most used spices in culinary preparations. This alkaloid shows anti-cancer activities in many different types of cancers. Piperine treatment remarkably suppressed both the androgen dependent and androgen independent tumor growth in nude mice model. Similarly, Piperine inhibited the metabolic activity of HRT-18 human rectal adenocarcinoma cells showing a cytostatic effect. Piperine inhibited cell cycle progression and induced apoptosis.

Rosemary (*Rosmarinus officinalis*), Family: Lamiaceae: Carnosic acid ($C_{20}H_{28}O_4$), carnosol ($C_{20}H_{26}O_4$) and rosmanol/rosmaquinone ($C_{20}H_{24}O_5$), rosmarinic acid/Labiatenic acid ($C_{18}H_{16}O_8$): The phytochemicals in this plant individually and collectively may have anti-cancer properties. Rosemary oil showed antitumor activity against breast cancer cells, and down-regulated estrogen-dependent-α and HER2 receptors. The effect of breast cancer chemotherapy was also significantly enhanced by super critical CO_2 extracts of Rosemary.

Resveratrol ($C_{14}H_{12}O_3$): Grapes: This is a natural phenol phytoalexin found in the skin of red grapes, peanuts, and other fruits. **Resveratrol** has proven anti-carcinogenesis effects in skin tumor in rats.

Salvicine: (*Salvia prionitis*): Salvicine is a diterpenoid quinone compound that inhibits DNA topoisomerase in breast cancer cell lines. **Salvicine** has also been found to have a profound cytotoxic effect on multidrug-resistant (MDR) cells. Chinese herbalists (TCM) prescribe it.

Sulforaphene/Glucoraphinin ($C_6H_9NOS_2$): *Brassica sp.*: Broccoli and other crucifers (*Brassica* sp): Sulforaphane is an organosulfur compound obtained from cruciferous vegetables such as broccoli, Brussels sprouts, and cabbages. The phytochemical **Glucoraphinin** present in these crucifers is converted to sulforaphine upon chewing or by action of enzymes in the intestines. The Sulforaphene has Chemopreventive activity and prevents mouth cancers. This chemical has cytotoxic properties with respect to other cancer cell lines such as those of prostate and colon.

Triterpinoids: cucurbitanes ($C_{30}H_{54}$), dammaranes ($C_{30}H_{54}$), ergostanes ($C_{28}H_{50}$), limonin ($C_{26}H_{30}O_8$), lupanes ($C_{30}H_{52}$), oleananes ($C_{30}H_{52}$), tirucallanes, ($C_{30}H_{54}$) and ursanes ($C_{30}H_{52}$): Triterpinoids are found in many plants such as Gourds/squashes/cumbers (*Cucurbita* sp), Indian penny wort (*Bacopa monnieri*), nightshade, Tomatillo, (*Solanum* sp), Ashwagandha (*Withania somnifera*), Citrus lemon (*Citrus limona*), Lupin (*Lupinus* sp), Fig (*Ficus carica*), blueberries, cranberries (*Vaccinium* sp) and evening primrose (*Mirabilis jalapa*). Triterpinoids have been shown to have pleiotropic anti-cancer effects on cancers in *in vitro* and *in vivo* models. They seem to exert their Chemopreventive and anti-cancer activities in pancreatic and breast cancer via enhancing apoptosis.

Triptolide, (C20H24O6): Thunder vine (*Tripterigium wilfordii*): Prescribed by Chinese herbalists for Rheumatoid arthritis and considered as anti-cell proliferative agent. However, it is extremely toxic and warnings about its use have been issued both by Chinese and UK regulatory agencies.

Ursolic Acid (C30H48O3): Four O Clock plant (*Mirabilis Jalapa*), Apple (*Pyrus malus*), Blue berries (*Vaccinium sp*) Cranberries (*Vaccinium oxycoccus*, *V. erythrocarpum*, *V. microcarpum*, *V, macrocarpum*), Hawthorn (*Crataegus sp.*), Lavender (*Lavendula sp.*), and rosemary (*Rosmarinus officinalis*): *In vitro* studies show that Ursolic acid inhibits various cancer cells.

Ineffective Anti-Cancer Formulations and Supplements

Various reputed cancer treatment and research centers have concluded that the following do not show any therapeutic value for treating Cancers. Some of these continue to be popular mainly because when people run out of options, they would try anything. The main sources for this list below are Memorial Sloan Kettering Institute, the Mayo Clinic, Web MD, Wikipedia, American Cancer Society, National Cancer Institute (USA), Cancer research UK.org. One or more of these prestigious organizations state that there is no evidence that any of the following have anti-cancer properties.

Aloe vera (Aloe): Extracts of this plant are sold as a cancer cure but according to Cancer Research UK, Aloe products do not prevent or treat cancer in humans. However, there are published reports attesting to a protective role for one of the phytochemicals known as Acemannan in cancer treatment.

Amygdalin (Laetrile): This glycoside extracted from Apricots as well as a synthetic form **laevo-mandelonitrile-beta-glucuronoside**, has been promoted as a cancer cure. It is ineffective and toxic. During 1962 and 1963, the Cancer Advisory Council examined more than 100 case histories given by various proponents and concluded that none supplied any evidence that **Laetrile** was effective against cancer. The Council also reviewed the California Medical Association's 1953 report on Laetrile, as well as a "new synthetic" Laetrile. The Council determined that the drug was "of no value in the diagnosis, treatment, alleviation or cure of cancer," it recommended that regulations be issued to ban the use of Laetrile and "substantially similar" agents for the treatment of cancer. Many other studies led to the conclusion that a systematic review that included all reports available through 2005 concluded that laetrile has NO beneficial effects for cancer patients. However, Laetrile continues to be promoted as a cancer cure in Mexico and many patients travel to Mexico hoping to get a cure.

Aveloz/Firestick plant/Pencil tree (*Euphorbia thirukalli*): As per the American Cancer Institute, Aveloz sap does not suppress Cancers but suppresses the immune system and promotes the development of certain types of cancer. It turns out that this plant is a hydrocarbon plant whose latex can be converted into gasoline.

Bach flower remedies: These are concoctions having the flower extracts of certain plants in a mixture of water and brandy. They are sold as remedies for diseases including Cancer without any supportive data.

Black cohosh (*Actaea racemosa*): Cancer Research UK states that available scientific evidence does not support claims that black cohosh is effective in treating or preventing cancer.

Cansema/Black salve (*Sanguinaria canadensis*): Extracts of Black Salve having **Sanguinarin** is marketed as a topical cream, paste, or poultice for treating skin conditions including melanoma. According to the American Cancer Society, there is no evidence that this is effective in treating cancer, and it can be harmful, causing burns and disfigurement.

Carctol: Carctol is an herbal oral supplement remedy that was widely reported to offer a cure for a range of cancers. The components are based on Ayurveda and one 560-mg capsule of Carctol contains extracts of *Hemidesmus indicus* roots, Tribulus *terrestris* seeds, Piper *cubeba* seeds, *Ammani vesicatoria* whole plant, *Lepidium sativum* seeds, *Blepharis edulis* seeds, Smilax *china* roots and *Rheum emodi* roots. The claim that Carctol is of any benefit to cancer patients is not supported by scientific evidence.

Cat's claw (*Uncaria tomentosa*): Extracts of this woody vine found in the tropical jungles of South and Central America is a part of some remedies for cancer and other diseases. The American cancer Society does not consider the products safe or effective in treating cancer.

Chaparral (*Larrea tridentata*): The leaf-extracts of this plant are sold as a treatment for cancer as well as several other diseases. The U.S. Food and Drug Administration and Health Canada have advised consumers against using products having chaparral due to safety concerns. Cancer Research UK state, "We don't recommend that you take chaparral to treat or prevent any type of cancer."

Chlorella (*Chlorella* sp): The dried cells of this single celled alga is promoted as a nutritional supplement as well as a treatment for cancer. This claim is not supported by the American Cancer Society and other cancer treatment centers.

Echinacea/Cone flower (Echinacea): Extracts from the ten species of *Echinacea* are marketed as a supplement to treat cancer as well as other diseases. Cancer Research UK has concluded, "There is no scientific evidence for these claims.

Ellagic acid (Oak bark, pomegranate, cranberries, black berries, strawberries, and walnut): This natural phenol found in berries, nuts and some trees are promoted as a treatment for Cancer. **Ellagic** acid has been identified by the U.S. Food and Drug Administration as a "fake cancer 'cure'.

Essiac tea: This is a blended herbal tea containing burdock (*Arctium* sp), American Indian Rhubarb (*Darnera peltata*), Sheep sorrel (*Rumex* sp) and slippery elm (Ulmus *rubra*) extracts that is promoted as a treatment for cancer. The U.S. Food and Drug Administration include Essiac in a list of "Fake Cancer 'Cures' that Consumers Should Avoid. The National Cancer Institute (USA), American Cancer Institute and the Cancer research UK do not endorse Essiac tea.

Ginseng (*Panax ginseng*): Root extracts of this plant are being marketed as a potential cancer treatment. However, the American Cancer Society concluded, "available scientific evidence does not support claims that ginseng is effective in preventing or treating cancer in humans".

Goldenseal (*Hydrastis canadensis*): The extracts from the rhizomes of this herb from the buttercup Ranunculaceae family is promoted as anti-catarrhal, anti-inflammatory, antiseptic, astringent, bitter tonic, laxative, anti-diabetic, and muscular stimulant. However, American Cancer Society does not support claims that goldenseal is effective in treating cancer or other diseases. Goldenseal can have toxic side effects, and high doses can cause death."

Gotu kola/Asian Indian pennywort (*Centella asiatica*): Supplements having leaf extracts of this swamp plant are promoted as cancer treatment. The American Cancer Society states "available scientific evidence does not support claims of its effectiveness for treating cancer or any other disease in humans".

Grapes (*vitis vinifera*): Consuming Grapefruits, grape-seed extract and drinking red wine is popularized for supposed anti-cancer effect. **Cyanidin** which is an extract of pigment from red berries such as grapes, blackberry, cranberry, raspberry, or apples and plums, red cabbage and red onion possesses antioxidant and radical-scavenging effects that may reduce the risk of cancer. It is reported to inhibit cell proliferation, and iNOS and COX-2 gene expression in colon cancer cells. However, The American Cancer Society does not endorse the anti-cancer properties of grapes.

Juice Plus: This is a proprietary trademark line of dietary supplements having concentrated fruit and vegetable juice extract. The effectiveness of this line of supplements was reviewed by the Sloan Kettering Institute, the University of California at Berkeley, and the Center for Science in the Public Interest. They have all questioned the efficacy of this concoction. The Memorial Sloan-Kettering Cancer Center have cautioned that while Juice Plus is being "aggressively promoted to cancer patients based on claims of antioxidant effects", the supplement should not be taken by patients because it can interfere with chemotherapy, nor should it be considered a substitute for fruit and vegetables. The American Cancer Society states that ""there is no convincing scientific evidence that extracted juices are healthier than whole foods.

Kālamegha/Bhūnimba/Green Chrayta (*Andrographis paniculata*): The extracts of this plant and its main constituent diterpine Andrographolide is being promoted as a dietary supplement for cancer prevention and cure. The Memorial Sloan-Kettering Cancer Center has said that there is no evidence that it helps prevent or cure cancer.

Kombucha: This is a fermented tea concoction originating from East Russia and Manchuria and now home brewed and sold in many health care's supplements out fits. It has no anti-cancer effects as claimed.

Mangosteen (*Garcinia mangostana*): The fruit of this plant that is native to Southeast Asia. It is promoted as a "super fruit" and in products such as XanGo Juice for treating a variety of human ailments. According to the American Cancer Society, "there is no reliable evidence that mangosteen juice, puree, or bark is effective as a treatment for cancer in humans".

Milk thistle (*Silybum marianum*): This biennial plant grows in many locations over the world. As per Cancer Research UK at milk, thistle extracts have no anti-cancer roles.

Mistletoe/Iscador (*Viscum album*): This is a parasitic plant related to sandal. Extracts of this plant are believed to control many diseases including cancer but there is no scientific basis for these claims.

Noni juice (*Morinda citrifolia*): The juice derived from the fruit of the Noni tree has been promoted as a cure for cancer. However, The American Cancer Society says, "There is no reliable clinical evidence that noni juice is effective in preventing or treating cancer or any other disease in humans".

Oleander (*Nerium oleander*): This is a poisonous garden plant that produces pink or white flowers. The seeds and fruit are Abortifacients and are lethal. According to the American Cancer Society, "even a small amount of oleander can cause death", and "the effectiveness of oleander has not been proven".

Pau d'arco/Lapacho tea/(*Handroanthus impetiginosus/Tabebuia*): The bark of the Lapacho tree is steeped in water to make a tea, which is consumed as a treatment for several diseases including cancer. The extracts are Abortifacients. According to the American Cancer Society, "available evidence from well-designed, controlled studies does not support this substance as an effective treatment for cancer in humans".

Saw palmetto (*Serenoa repens*): Extracts of this palm tree is promoted as a treatment for prostate cancer. The American Cancer society considers that these claims are unsubstantiated.

Snakeroot (*Rauvolfia serpentina*): The Indian snakeroot plant has the alkaloid **reserpine** that is an approved treatment for hypertension but claims for its effectiveness in treating cancers is not substantiated. It has side effects that can be fatal.

Seasilver: This proprietary product having pau d'arco, cranberry, aloe and other ingredients is an expensive dietary supplement that is claimed to help health and treatment of many diseases including Cancer. The US Food and Drug administration fined the manufacturers for making unsubstantiated claims. The Memorial Cancer Institute also could not verify any of these medicinal claims

Soursop/Graviola (*Annona muricata*): The fruit and leaf extracts of this plant are promoted as a cure and treatment for cancer. The fruits and leaves have a potent neurotoxin and some studies link consumption of the fruit to atypical Parkinson's disease. The Memorial Sloan Kettering Institute, Cancer research UK and Federal trade commission of USA do not endorse claims about Soursop as a treatment choice for cancer. It is an ineffective treatment heavily promoted on the internet

Ukrain/Celandine: *This is a trademark product having extracts from* Chelidonium *majus, a plant in the poppy (Papavaraceae) family. The drug that is marketed by an Austrian company is promoted for its health-giving powers and its ability to treat cancer.* However, according to the American Cancer Society, "available scientific evidence does not support claims that celandine is effective in treating cancer in humans". The Memorial Sloan Kettering Cancer Institute also does not endorse this product. The drug is illegal in the USA.

Venus flytrap (*Dionaea muscipula*): This is – a carnivorous plant found in the wetlands of the Eastern US seaboard. The extracts of this plant are marketed as a treatment for a variety of human ailments including skin cancer. According to the American Cancer Society, "available scientific evidence does not support claims that extract from the Venus flytrap plant is effective in treating skin cancer or any other type of cancer".

Wheatgrass (common wheat-*Triticum vulgaris*): Wheat grass juice made from the young first leaves of the common wheat or from wheat sprouts is another juice product that many juice bars serve and is consumed as a food supplement for prevention and treatment of cancer. According to the American Cancer Society, although some wheatgrass champions claim it can "shrink" cancer tumors, "available scientific evidence does not support the idea that wheatgrass or the wheatgrass diet can cure or prevent disease".

Wild yam (or Chinese yam – types of yam): The roots of these are made into creams and dietary supplements that are promoted for a variety of medicinal purposes, including cancer prevention. The American Cancer Society says of these products, "available scientific evidence does not support claims that they are safe or effective.

Comments on Herbals as Cure for Cancer

While many plants have been found as sources for anti-cancer agents, there is no evidence that shows that any herbal treatment *per se* can control cancers. However, herbals may play a role in complementing modern medical treatments as well as function as chemopreventive agents. The search for anti-cancer agents in plants should continue. It must also be pointed out that in many instances' treatment based on whole plant extracts which have many phytochemicals have been condemned based on tests on a single part of the plant. Herbalists would argue that they are not recommending a single chemical in a plant product but the entire or specific parts of a plant having major and minor constituents that collectively function as a medicinal.

Table 8: Complete list of different types of cancers, adapted from National Cancer Institute (USA): https://www.cancer.gov/types#y

Acute Lymphoblastic Leukemia (ALL)	Central Nervous System
Acute Myeloid Leukemia (AML)	Embryonal Tumors, Childhood (Brain Cancer)
Adolescents, Cancer in	
Adrenocortical Carcinoma	Germ Cell Tumor, Childhood (Brain Cancer)
Childhood Adrenocortical	
AIDS-Related Cancers	Primary CNS Lymphoma
Kaposi Sarcoma (Soft Tissue Sarcoma)	Cervical Cancer
AIDS-Related Lymphoma (Lymphoma)	Childhood Cervical Cancer -
Primary CNS Lymphoma (Lymphoma)	Childhood Cancers
Anal Cancer	Cancers of Childhood, Unusual
Appendix Cancer - see Gastrointestinal Carcinoid Tumors	Cholangiocarcinoma - Bile Duct Cancer
	Chordoma, Childhood
Astrocytoma, Childhood (Brain Cancer)	Chronic Lymphocytic Leukemia (CLL)
Atypical Teratoid/Rhabdoid Tumor, Childhood, Central Nervous System (Brain Cancer)	Chronic Mylogenous Leukemia (CML)
	Chronic Myeloproliferative Neoplasms
	Colorectal Cancer
Basal Cell Carcinoma of the Skin	Childhood Colorectal Cancer
Bile Duct Cancer	Craniopharyngioma, Childhood (Brain Cancer)
Bladder Cancer	
Childhood Bladder Cancer -	Cutaneous T-Cell Lymphoma
Bone Cancer (includes Ewing Sarcoma and Osteosarcoma and Malignant Fibrous Histiocytoma)	Ductal Carcinoma in Situ (DCIS)- Breast Cancer
	Embryonal Tumors, Central Nervous System, Childhood (Brain Cancer)
Brain Tumors	
Breast Cancer	Endometrial Cancer (Uterine Cancer)
Childhood Breast Cancer	Ependymoma, Childhood (Brain Cancer)
Bronchial Tumors	Esophageal Cancer
Burkitts Lymphoma	Childhood Esophageal Cancer - see Unusual Cancers of Childhood
Carcinoid Tumor (Gastrointestinal)	
Childhood Carcinoid Tumors -	Esthesioneuroblastoma (Head and Neck Cancer)
Carcinoma of Unknown Primary	
Childhood Carcinoma of Unknown Primary -	Ewing Sarcoma (Bone Cancer)
Cardiac (Heart) Tumors,	Extra cranial Germ Cell Tumor, Childhood

Extra gonadal Germ Cell Tumor	Islet Cell Tumors, Pancreatic Neuroendocrine Tumors
Eye Cancer	
Childhood Intraocular Melanoma - see Unusual Cancers of Childhood	Kaposi Sarcoma (Soft Tissue Sarcoma)
	Kidney (Renal Cell) Cancer
Intraocular Melanoma	Langerhans Cell Histiocytosis
Retinoblastoma	Laryngeal Cancer (Head and Neck Cancer)
Fallopian Tube Cancer	Leukemia
Fibrous Histiocytoma of Bone, Malignant, and Osteosarcoma	Lip and Oral Cavity Cancer (Head and Neck Cancer)
Gallbladder Cancer	Liver Cancer
Gastric (Stomach) Cancer	Lung Cancer (Non-Small Cell and Small Cell)
Childhood Gastric (Stomach) Cancer - see Unusual Cancers of Childhood	
	Childhood Lung Cancer
Gastrointestinal Carcinoid Tumor	Lymphoma
Gastrointestinal Stromal Tumors (GIST) (Soft Tissue Sarcoma)	Male Breast Cancer
	Malignant Fibrous Histiocytoma of Bone and Osteosarcoma
Childhood Gastrointestinal Stromal Tumors	
Germ Cell Tumors	Melanoma
Childhood Central Nervous System Germ Cell Tumors (Brain Cancer)	Childhood Melanoma
	Melanoma, Intraocular (Eye)
Childhood Extra cranial Germ Cell Tumors	Childhood Intraocular Melanoma
Extra gonadal Germ Cell Tumors	Merkel Cell Carcinoma (Skin Cancer)
Ovarian Germ Cell Tumors	Mesothelioma, Malignant
Testicular Cancer	Childhood Mesothelioma -
Gestational Trophoblastic Disease	Metastatic Cancer
Hairy Cell Leukemia	Metastatic Squamous Neck Cancer with Occult Primary (Head and Neck Cancer)
Head and Neck Cancer	
Heart Tumors, Childhood	Midline Tract Carcinoma with NUT Gene Changes
Hepatocellular (Liver) Cancer	
Histiocytosis, Langerhans Cell	Mouth Cancer (Head and Neck Cancer)
Hodgkin Lymphoma	Multiple Endocrine Neoplasia Syndromes
Hypo pharyngeal Cancer (Head and Neck Cancer)	Multiple Myeloma/Plasma Cell Neoplasms
	Mycosis Fungoides (Lymphoma)
Intraocular Melanoma	Myelodysplastic Syndromes, Myelodysplastic/Myeloproliferative Neoplasms
Childhood Intraocular Melanoma - see Unusual Cancers of Childhood	
	Myelogenous Leukemia, Chronic (CML)

Continued...

Myeloid Leukemia, Acute (AML)	Prostate Cancer
Myeloproliferative Neoplasms, Chronic	Rectal Cancer
Nasal Cavity and Para nasal Sinus Cancer (Head and Neck Cancer)	Recurrent Cancer
	Renal Cell (Kidney) Cancer
Nasopharyngeal Cancer (Head and Neck Cancer)	Retinoblastoma
	Rhabdomyosarcoma
Neuroblastoma	Salivary Gland Cancer (Head and Neck Cancer)
Non-Hodgkin Lymphoma	
Non-Small Cell Lung Cancer	Sarcoma
Oral Cancer, Lip and Oral Cavity Cancer and Oropharyngeal Cancer (Head and Neck Cancer)	Rhabdomyosarcoma (Soft Tissue Sarcoma)
	Childhood Vascular Tumors (Soft Tissue Sarcoma)
Osteosarcoma and Malignant Fibrous Histiocytoma of Bone	Ewing Sarcoma (Bone Cancer)
	Kaposi Sarcoma (Soft Tissue Sarcoma)
Ovarian Cancer-	Osteosarcoma (Bone Cancer)
Ovarian Cancer-childhood	Soft Tissue Sarcoma
Pancreatic Cancer	Uterine Sarcoma
Pancreatic Cancer - Childhood	Sézary Syndrome (Lymphoma)
Pancreatic Neuroendocrine Tumors (Islet Cell Tumors)	Skin Cancer
	Childhood Skin Cancer
Papillomatosis (Childhood Laryngeal)	Small Cell Lung Cancer
Paraganglioma	Small Intestine Cancer
Childhood Paraganglioma - see Unusual Cancers of Childhood	Soft Tissue Sarcoma
	Squamous Cell Carcinoma of the Skin
Paranasal Sinus and Nasal Cavity Cancer (Head and Neck Cancer)	Squamous Neck Cancer with Occult Primary, Metastatic (Head and Neck Cancer)
Parathyroid Cancer	
Penile Cancer	
Pharyngeal Cancer (Head and Neck Cancer)	Stomach (Gastric) Cancer
Pheochromocytoma	Childhood Stomach (Gastric) Cancer
Pituitary Tumor	T-Cell Lymphoma, Cutaneous -
Plasma Cell Neoplasm/Multiple Myeloma	Testicular Cancer
Pleuropulmonary Blastoma	Childhood Testicular Cancer -
Pregnancy and Breast Cancer	Throat Cancer (Head and Neck Cancer)
Primary Central Nervous System (CNS) Lymphoma	Nasopharyngeal Cancer
	Oropharyngeal Cancer
Primary Peritoneal Cancer	Hypo pharyngeal Cancer

Thymoma and Thymic Carcinoma	Urethral Cancer
Thyroid Cancer	Uterine Cancer, Endometrial
Transitional Cell Cancer of the Renal Pelvis and Ureter (Kidney (Renal Cell) Cancer)	Uterine Sarcoma
	Vaginal Cancer
Unknown Primary, Carcinoma of	Childhood Vaginal Cancer
Childhood Cancer of Unknown Primary -	Vascular Tumors (Soft Tissue Sarcoma)
Unusual Cancers of Childhood	Vulvar Cancer
Ureter and Renal Pelvis, Transitional Cell Cancer (Kidney (Renal Cell) Cancer	Wilms Tumor and Other Childhood Kidney tumors

References

https://www.cancerresearchuk.org/about-cancer/cancer-in-general/treatment/complementary-alternative-therapies/individual-therapies/turmeric.

https://www.mskcc.org/cancer-care/diagnosis-treatment/symptom-management/integrative-medicine/herbs.

Arpita Roy, Shruti Ahuja, Navneeta Bharadvaja (2017): A Review on Medicinal Plants against Cancer J Plant Sci Agric Res. Vol. 2 No. 1: 008.

Aung HH, Wang CZ, Ni M, Fishbein A, Mehendale SR, Xie JT, Shoyama CY, Yuan CS. (2007): Crocin from Crocus sativus possesses significant anti-proliferation effects on human colorectal cancer cells. Exp Oncol. 2007, 29(3): 175–180. [PMC free article]

Bachmeier BE, Mirisola V, Romeo F, Generoso L, Esposito A, Dell'eva R, Blengio F, Killian PH, Albini A, Pfeffer U. (2010): Reference profile correlation reveals estrogen-like trancriptional activity of Curcumin. Cell Physiol Biochem. 2010, 26(3): 471–482.

Bathaie SZ, Mousavi SZ. (2010): New applications and mechanisms of action of saffron and its important ingredients. Crit Rev Food Sci Nutr. 2010, 50(8): 761–786.

Calderon-Montano JM, Burgos-Moron E, Perez-Guerrero C, Lopez-Lazaro M. (2011): A review on the dietary flavonoid kaempferol. Mini Rev Med Chem. 2011, 11(4): 298–344.

Canene-Adams K, Lindshield BL, Wang S, Jeffery EH, Clinton SK, Erdman JW., Jr. (2007): Combinations of tomato and broccoli enhance antitumor activity in dunning r3327-h prostate adenocarcinomas. Cancer Res. 2007, 67(2): 836–843.

Claudio Festuccia, Alessandro Colapietro, Andrea Mancini, and Annamaria D'alessandro (2019): Crocetin and Crocin from Saffron in Cancer Chemotherapy and Chemoprevention, Anti-Cancer Agents in Medicinal Chemistry, DOI: 10.2174/1871520619666181231112453

Clere N, Faure S, Martinez MC, Andriantsitohaina R. (2011): Anticancer properties of flavonoids: roles in various stages of carcinogenesis. Cardiovasc Hematol Agents Med Chem. 2011, 9(2): 62–77.

Epstein J, Sanderson IR, Macdonald TT. (2010): Curcumin as a therapeutic agent: the evidence from in vitro, animal, and human studies. Br J Nutr. 2010, 103(11): 1545–1557.

Friedlander, M., Kapulnik, Y., Koltai, H. (2015): Plant derived substances with anti-cancer activity: From folklore to practice. Front. Plant Sci. 2015; 6: 799. Doi: 10.3389/fpls.2015.00799. [PMC free article]

Greenwell M, Rahman PKSM (2015) Medicinal Plants: Their Use in Anticancer Treatment. International journal of pharmaceutical sciences and research 6: 4103-4112.

Gutheil WG, Reed G, Ray A, Dhar A. (2011): Crocetin: An Agent Derived from Saffron for Prevention and Therapy for Cancer. Curr Pharm Biotechnol. 2011[PMC free article]

Hu R, Khor TO, Shen G, Jeong WS, Hebbar V, Chen C, Xu C, Reddy B, Chada K, Kong AN. (2006): Cancer chemoprevention of intestinal polyposis in ApcMin/+ mice by sulforaphane, a natural product derived from cruciferous vegetable. Carcinogenesis. 2006, 27(10): 2038–2046.

Hu Wang, Tin Oo Khor, Limin Shu, Zhengyuen Su, Francisco Fuentes, Jong-Hun Lee, and Ah-Ng Tony Kong (2012): Plants Against Cancer: A Review on Natural Phytochemicals in Preventing and Treating Cancers and Their Druggability. Anticancer Agents Med Chem. 2012 Dec, 12(10): 1281–1305. PMCID: PMC4017674, NIHMSID: NIHMS578519, PMID: 22583408.

Huang WY, Cai YZ, Zhang Y. (2010): Natural phenolic compounds from medicinal herbs and dietary plants: potential use for cancer prevention. Nutr Cancer. 2010, 62(1): 1–20.

Jang M, Cai L, Udeani GO, Slowing KV, Thomas CF, Beecher CW, Fong HH, Farnsworth NR, Kinghorn AD, Mehta RG, Moon RC, Pezzuto JM.(1997): Cancer chemopreventive activity of resveratrol, a natural product derived from grapes. Science. 1997, 275(5297): 218–220.

Jeong SJ, Koh W, Lee EO, Lee HJ, Lee HJ, Bae H, Lu J, Kim SH. (2011): Antiangiogenic phytochemicals and medicinal herbs. Phytother Res. 2011, 25(1): 1–10.

Karikas GA. (2010): Anticancer and chemopreventing natural products: some biochemical and therapeutic aspects. J Buon. 2010; 15(4): 627–638., Biochem., 2010; 48(7): 621–626.

Kayande N, Patel R (2016) Review on: Indian Medicinal plants having anticancer property. Pharma Tutor 4: 25-28.

Kim KH, Back JH, Zhu Y, Arbesman J, Athar M, Kopelovich L, Kim AL, Bickers DR. (2011): Resveratrol targets transforming growth factor-beta2 signaling to block UV-induced tumor progression. J Invest Dermatol. 2011, 131(1): 195–202.

Leonardi T, Vanamala J, Taddeo SS, Davidson LA, Murphy ME, Patil BS, Wang N, Carroll RJ, Chapkin RS, Lupton JR, Turner ND. (2010): Apigenin and naringenin suppress colon carcinogenesis through the aberrant crypt stage in azoxymethane-treated rats. Exp Biol Med (Maywood) 2010; 235(6): 710–717. [PMC free article].

Li F., Li S., Li H.B., Deng G.F., Ling W.H., Xu X.R. (2013): Antiproliferative activities of tea and herbal infusions. Food Funct. 2013; 4: 530–538, doi: 10.1039/c2fo30252g.

Li Y, Zhang T, Korkaya H, Liu S, Lee HF, Newman B, Yu Y, Clouthier SG, Schwartz SJ, Wicha MS, Sun D.(2010) Sulforaphane, a dietary component of broccoli/broccoli sprouts, inhibits breast cancer stem cells. Clin Cancer Res. 2010; 16(9): 2580–2590. [PMC free article].

Li, Sha Li, Shu-Key Gan, Ren-You Song, Feng-Lin et al (2013): Phenolic contents of infusions from 223 medicinal plants. Ind. Crop Prod., 2013; 51: 289–298. Doi: 10.1016/j.indcrop.2013.09.017.

Li-Weber M. (2013): Targeting apoptosis pathways in cancer by Chinese medicine. Cancer Lett. 2013 May 28, 332(2): 304-12. Doi: 10.1016/j.canlet.2010.07.015. Epub 2010 Aug 3.

Mehta RG, Murillo G, Naithani R, Peng X. (2010), Cancer chemoprevention by natural products: how far have we come? Pharm Res. 2010, 27(6): 950–961.

Nothlings U, Murphy SP, Wilkens LR, Henderson BE, Kolonel LN. (2007): Flavonols and pancreatic cancer risk: the multiethnic cohort study. Am J Epidemiol. 2007, 166(8): 924–931.

Park YJ, Wen J, Bang S, Park SW, Song SY. (2010): [6]-Gingerol induces cell cycle arrest Oyagbemi AA, Saba AB, Azeez OI. Molecular targets of [6]-gingerol: Its potential roles in cancer chemoprevention. Biofactors. 2010, 36(3): 169–178.

Raynal NJ, Momparler L, Charbonneau M, Momparler RL. (2008): Antileukemic activity of genistein, a major isoflavone present in soy products. J Nat Prod. 2008, 71(1): 3–7.

Rubio L., Motilva M.J., Romero M.P. (2013): Recent advances in biologically active compounds in herbs and spices: A review of the most effective antioxidant and anti-inflammatory active principles. Crit. Rev. Food Sci. Nutr. 2013; 53: 943–953, doi: 10.1080/10408398.2011.574802.

Sanchez-Fidalgo S, Cardeno A, Villegas I, Talero E, de la Lastra CA.92010): Dietary supplementation of resveratrol attenuates chronic colonic inflammation in mice. Eur J Pharmacol. 2010, 633(1-3): 78–84.

Sarkar FH, Li Y, Wang Z, Padhye S. (2010), Lesson learned from nature for the development of novel anti-cancer agents: implication of isoflavone, curcumin, and their synthetic analogs. Curr Pharm Des. 2010, 16(16): 1801–1812.

Scheckel KA, Degner SC, Romagnolo DF. (2008): Rosmarinic acid antagonizes activator protein-1-dependent activation of cyclooxygenase-2 expression in human cancer and nonmalignant cell lines. J Nutr. 2008, 138(11): 2098–2105.

Shanmugam M.K., Rane G., Kanchi M.M., Arfuso F., Chinnathambi A., Zayed M.E., Alharbi S.A., Tan B., Kumar A.P., Sethi G. (2015), The multifaceted role of curcumin in cancer prevention and treatment, Molecules, 2015,20: 2728–2769. Doi: 10.3390/molecules20022728. [PMC free article].

Stan SD, Singh SV, Brand RE., (2010): Chemoprevention strategies for pancreatic cancer. Nat Rev Gastroenterol Hepatol. 2010, 7(6): 347–356.

Turktekin M, Konac E, Onen HI, Alp E, Yilmaz A, Menevse S.92011), Evaluation of the Effects of the Flavonoid Apigenin on Apoptotic Pathway Gene Expression on the Colon Cancer Cell Line (HT29) J Med Food. 2011.

CHAPTER 6

Herbals for Cardiovascular System

The human cardiovascular system consists of the four-chambered heart with the right atrium and ventricle and the left atrium and ventricle separated by a septum and the network of arteries and veins, which distribute the oxygenated blood to different parts of the body and return the de-oxygenated blood for oxygenation to the heart and from there to lungs and back to heart after oxygenation in lungs. Thus, de-oxygenated blood from the body tissues circulates into the right atrium by the veins. From the right atrium, the blood moves into the right ventricle that then pumps the blood for oxygenation to the lungs. From the lungs, the blood returns to the left atrium and then on to the left ventricle which then pumps the oxygenated blood to the rest of the body. Thus, the heart pumps blood to different parts of the body through a network of arteries, arterioles, and veins with veinlets. Any weakening of the heart muscle and blockages anywhere in this network will result in cardiovascular disease. These cardiovascular diseases include: 1. Stroke 2. Heart failure 3. High Blood pressure (Hypertension) 4. Low Blood Pressure (Hypotension) 5. High pulse (Tachycardia) 6. Low pulse (Bradycardia) 7. Peripheral circulatory diseases including Venous insufficiency (veins not pumping back the de-oxygenated blood to the heart) 8. Various circulatory disorders including aneurism 9. Blood clots 9. All other factors that prevent proper functioning of the circulatory system.

Cardiovascular Diseases Result From

a. Inherited congenital conditions that can sometimes be corrected by surgical intervention. Thus, inherited valve defects can be repaired by artificial valves and certain surgical procedures. Other inherited conditions cannot be repaired except through transplantation in admissible cases. Stenosis, regurgitation, and mitral valve prolapse are often inherited valve conditions but sometimes result due to viral and bacterial infections.

b. Atherosclerosis/Stroke: The heart vessals become thickened due to buildup of plaque. The blood vessels become narrow and blood supply to the heart or brain or other parts of the body is impaired or even cut off. This can result in weakening of the heart muscle and hence poor circulation and a general condition known as heart failure results. Alternatively, an ischemic stroke can occur when blood supply to brain is reduced or cut off thereby impairing speech, hearing, partial paralysis, facial movements etc. Another type of stroke known as hemorrhagic stroke occurs when blood vessals break in the brain usually due to high blood pressure. Other conditions include arrhythmias that occur when the heart beats too slowly (bradycardia) or too fast (tachycardia).

Many of these conditions can be controlled by proper diet, exercise, meditation, and relaxation. Diabetic conditions potentiate heart disease and hence control of diabetes is important. Good Cardiovascular health depends upon good dietary habits by excluding high fat, high sugar, high salt foods by substituting with oils rich in omega-3 fats, fruit sugar and low salt foods. The diet should preferably include high fiber grains, green and colored vegetables like broccoli, spinach, carrots, beans, okra etc. Regular exercise, meditation and inclusion of antioxidants and other useful phytochemicals from herbs in the diet are also recommended. Heart diseases stem from blockage of the arteries and veins through excessive deposition of cholesterol resulting in plaque formation. The risk factor for atherosclerosis is high fat and cholesterol diets, smoking, lack of exercise, high blood pressure, obesity, diabetes, and certain genetic factors. The best options for preventing cardiovascular disease are by prevention rather than by medication although certain medications do help in preventing further deterioration of the arterial and cardiovascular system.

The following plants through their plant products could be important for helping to ensure good cardiac health. Plant foods rich in phytochemicals and having low glycemic index plus high fiber are good for good health.

A list and descriptions of plants useful in cardiovascular health is given below. This list does not include food plants.

Ammi/Khella/toothpick plant/Bishops weed (*Ammi visnaga*): More detailed descriptions in the chapter on Skin. The phytochemical Khellin from this plant is known to control heart arrhythmia but is not as efficient or safe as many of the new medications like Amiodarone.

Arjuna: (*Terminalia Arjuna*) Kingdom: Plantae Group: Angiospermae Eudicots; Order: Myrtales Family: Combretaceae. Other names: Arjuna- Sanskrit; Marudha maram-Siddha and Tamil, Neer maruthu in Malayalam, kumbuk in Sinhallese. Related species are T. *bellerica* and T. *chebula*.

Arjuna is a large tree with leaves and branches at the canopy. The leaves droop down. Arjuna extracts stimulates the heart. Its bark is used in Ayurveda for cardio protective treatments. It is also used for treating wounds and ulcers. The role of Arjuna in cardio protection is not clear but nevertheless it continues to be prescribed for heart ailments by Ayurvedic physicians. The Triterpene glycosides **arjunetin, arjunetoside, arjunaphthanoloside** together with oleanolic and **arjunic acids, terminic acid**, strong antioxidants, flavones, tannins and oligomeric proanthocyanidins have been isolated from this plant.

Bilberry/whortleberry: (*Vaccinium myrtillus*) Kingdom: Plantae, Phylum: Angiospermae, Group: Dicotyledonae; Family: Ericaceae.

Bilberry is closely related to blue berry. Plants are shrubs growing wild in Europe and in North America. Leaves are glabrous, bright green, elliptic or ovate with serrated margins. Flowers are pink green or cream colored. Fruits are berries, which are purple, black, or bluish black. Both the berry skin and pulp are pigmented. They have anthocyanin. Fruits are a rich source of **flavonoids** with high antioxidant activity. Eating the fruits is considered to confer good circulatory effects, reduce oxidative stress, and generate antiinflammatory activities. Some studies also show that it helps to control diabetes and hypertension-induced retinopathy. Dried fruits are recommended for controlling acute diahorrea. There are no known adverse effects due to consumption of bilberry fruits.

Blue berries (*Vaccinium sp.*): Kingdom: Plantae, Clade: Angiospermae, Eudicots, Order: Ericales, Family: Ericaceae, Genus: Vaccinium.

Blue berry plants are perennial shrubs growing in the cold climes of North America but have been introduced in Europe, Australia, South America, and temperate regions of Asia. The plants fall into two general categories namely low bush varieties (V. *angustifolium*, 4-5 inch tall) which produce small globose fruits and taller large bush varieties (> 10 ft tall, V. *corymbosum*) which are the general choice among growers. Both varieties produce bell-shaped pink flowers and dark blue berries. The berries are coated with wax and so they appear as if they have been sprayed with a whitish protective agent. Fruits are extremely rich in **flavones, anthocyanins,** and antioxidants, which make the fruits ideal for conferring many health benefits including cardio-protection.

Citrus (*Citrus sp.*): Kingdom: Plantae, Eudicots, Order: Sapindales, Family: Rutaceae. Oranges, lemons. lime, grapefruit, pomelo, and kumquats are all examples of citrus fruits. The rinds and fruit juice of oranges and lemons have bioflavonoids including **Hesperidin ($C_{28}H_{34}O_{15}$) and Diosmin ($C_{28}H_{32}O_{15}$)**. Both these phytochemicals are used in treatment of venous insufficiency which results in spider veins, venous reflux disease and peripheral circulation. Hesperidin powder along with Diosmin in the form of capsules is used. It along with Diosmin is also used in creams to control leg cramps and swelling of legs. The orange oil also has **furano coumarins** which could cause photo dermatitis if skin smeared with the oil is exposed to light.

Chinese Salvia/Danshen/Chinese red Sage (*Salvia miltiorrhiza*); Kingdom: Plantae, Eudicots, Order: Lamiales, Family: Lamiaceae.

The plants are perennial herbs grown in China and japan. The roots and rhizomes are used in TCM for cardiovascular health, but the evidence is not conclusive. It has **Salvianolic acid** as well as other phytochemicals. Although used in Chinese medicine, the plant parts are used to treat Chest pain (angina), high cholesterol, high blood pressure and ischemic stroke. It increased the blood thinning effects of Warfarin and hence people on blood thinners should use care if using this herbal.

Cinnamon-Ceylon/Sri Lanka (*Cinnamon Cinnamomum verum*/C. *Zeylanicum*): Kingdom: Plantae; Group: Angiospermae Eudicots, Order: Laurales, Family: Lauraceae. Other names: tamalapatram-Sanskrit; Quinnamon-Hebrew.

The plants which are natives of Sri Lanka and Kerala state of India are trees which grow to a height of about 50 ft. Leaves are oval about 37 inches long. Flowers are arranged in panicles and are green in color. The fruit is a drupe. Cinnamon bark pieces or powder is used to flavor foods. Indian Ayurvedic healers used cinnamon for treating digestive and menstrual complaints. The herb is also recommended for improving peripheral blood flow. However, recent studies show that it has insulin like properties by stabilizing blood sugar levels in people with mild type 2 diabetes. It has about **60 different aromatic** compounds including **cinnamaldehyde and coumarin. Usage:** Adding powdered cinnamon to beverages and foods are said to reduce blood sugar. Capsules are also available. **Safety:** It can increase bleeding in persons who are already taking blood thinners.

Cocoa/chocolate (*Theobroma cacao*): Kingdom: Plantae, Clade: Angiospermae, Eudicots, Order, Malvales, Family: Malvaceae, Genus and species: Theobroma cacao

Cocoa trees are small evergreen trees with fruiting pods having cocoa seeds. The seeds are processed into chocolate. Seeds are rich in **flavanoids** and antioxidants. Cocoa and chocolate products are helpful in protecting the cardiovascular system. Cocoa and chocolate help reduce inflammation, reduce risks of stroke, and improve cardio circulation.

Fox glove Lady's glove/Hritpatri, Tilapushpi (*Digitalis purpurea*): Kingdom: Plantae Group: Angiospermae Dicotyledonae, Order: Lamiales, Family: Plantiginaceae.

Digitalis is an herbaceous biennial or short-lived perennial plant. The leaves are spirally arranged simple 10–35 cm long and 5–12 cm broad and are covered with gray, white pubescent and glandular hairs imparting a woolly texture. The foliage forms a tight rosette at ground level in the first year. The flowers are arranged in spikes. They vary in color from white to pink rose or yellow violet to purple. The fruit is a capsule, which splits open at maturity to release the many tiny 0.10.2 mm seeds.

The leaves have two cardiac glycosides namely **digoxin and digoxygenin**. Extracts of leaves and pure glycosides are used for treating heart conditions.

Garlic (*Allium sativum*): Kingdom: Plantae Group: Angiospermae, Monocotyledonae, Order: Asperagales, Family: Amaryllidaceae, Related species are A. *longicuspis*, A. *vineale*, A. *oleraceum*, A. *canadense*, and A. *ampeloprasum*.

Onion, Shallot, Leek, Rakkyo, and Chive are close relatives of garlic. *Allium sativum* is a bulbous plant growing up to 4 ft. in height. It produces hermaphrodite flowers.

Garlic is native to central Asia and has long been a staple in the Mediterranean region as well as a seasoning agent in Asia, Africa, and Europe. Currently, China followed by India are the largest growers and producers. It is a universally used herb in the cuisines of Italy, France, China, India, and many other nations. Garlic yields **allicin** an antibiotic and the sulfur having compounds **alliin, ajoene, diallyl polysulfides, vinyldithiins, Allylcysteine** and enzymes, B vitamins, proteins, minerals, Saponins and flavonoids. Garlic has long been used to help reduce serum cholesterol levels and in general to improve heart health. However, while some human studies found garlic supplementation to produce small reductions in blood cholesterol, a National Center for complementary and Integrative Health (NCCIH) funded study found no effect. According to a meta-analysis from 2009, garlic has no beneficial effect on serum cholesterol levels either in healthy people or in people with hypercholesterolemia. Garlic is often stored in Oil before use, but such stored garlic is prone to allow growth of the deadly botulism bacteria *clostridium botulinum*. Although Garlic has antimicrobial properties, the skin allows growth of molds and hence garlic that is soft and soggy should not be eaten since it may have microbial toxins. Some People suffer from Garlic allergies resulting in irritable bowel syndrome, diahorrea, gastric distress and sometimes anaphylaxis. Applying garlic juice on skin can result in burns, especially in children.

Ginkgoa (*Ginkgoa biloba*) Kingdom: Plantae, Clade: Gymnospermae, Ginkgophyta, order: Ginkoales, Family: Ginkoaceae

Ginkgo is a gymnosperm since the seeds are not protected by an ovular wall. They have naked seeds and the life cycle includes stages reminiscent of both gymnosperms and the pteridipyhta. Ginkgo plants are large deciduous trees about hundred and twenty feet high with fan shaped to lobed leaves with prominent fourth veins. The leaves are borne in clusters or as alternate leaves. There are separate male and female trees. The male trees produce pollen in microsporophyll and the females produce mega sporophylls having ovules. Fertilization results in the production of fruits having a single ovoid seed. Ginkgo leaves have **glycosides and terpene lactones** which have antioxidant and antiinflammatory properties. **Usage:** Ginkgo leaf extracts, tea and tinctures are prescribed by practitioners of Chinese traditional medicine for improving cognitive functions and improved blood circulation. However, studies conducted in the US do not substantiate the beneficial effects of ginkgo products in improving mental health and cognitive functions. **Warning:** Gingko may have undesirable side effects in individuals taking blood-thinning agents since it may increase bleeding.

Grapes: (*Vitis vinifera*) Kingdom: Plantae; Group; Angiospermae, Class: Eudicots, Order: Vitales, Family: Vitaceae. Chromosome number 2n=38

Grape juice, grape seed extracts and grape wine have resveratrol. Various studies have shown that resveratrol helps to keep good cardiac health. The seed oil has unsaturated fatty acids that help to lower LDL cholesterol.

Guggul/Indian bdellium tree/Mukul/(*Commiphora wightii*): Kingdom: Plantae Group: Angiosperms, Eudicots, Order: Sapindales, Family: Burseraceae Synonyms: *Commiphora mukul, Commiphora roxburghii*

Gugul is a small tree that grows to a height of about 8 to 10 feet. It is a native northern Africa, Central Asia, and northern India. It produces simple trifoliate leaves with irregular serrated edges. The plants may be bisexual with male flowers on others female flowers. Flowers are red or pink with four small petals. Fruits are small and red in color. The resin from the tree is used in Ayurvedic medicine for treatment of cardiovascular conditions as well as an antiinflammatory drug. Some studies show that it may be useful for treatment of arthritis, Crohn's disease, diabetes, obesity, and psoriasis. The cardio protective effects were not confirmed by studies conducted in the United States although other studies in India show that it has powerful cholesterol lowering properties. The active ingredient in Gugul lipid is **gugulsterone**. Usage: Guggul resin and powders in tablet form.

Hawthorn/Mayflower *(Crataegus laevigata)*: Kingdom: Plantae; Group; Angiospermae, Eudicots, order: Rosales, Family: Rosaceae, C. *pinnatifida* C. *monogyna*.

Plants are large shrubs or small trees with spiny trunks and long leaves. Flowers are small with five petals and are white or pink in color. The fruit is a dark red pome 6–10 mm diameter slightly broader than long having 2–3 nutlets. Hawthorn can improve heart health. **Usage:** The leaves and flowers are steeped in water overnight and drunk as a tea or a tincture of the leaves and flowers are prepared for treatment.

Hibiscus *(Hibiscus subdariffa, Hibiscus Roselle)*: Kingdom: Plantae Group: Angiospermae; Dicotyledonae; Order: Malvales Family: Malvaceae. Other related species: *H. Rosasinensis, H. mutablis, H. tiliaceus, H. hirta, H. hispidissimus, H. arnottianus*.

Plants grow in warm climates mostly subtropical and tropical countries although the plants of this genus are also grown in temperate countries as houseplants or grown in the summer because of their spectacular flowers. Irrespective of species, plants of this genus are shrubs, which produce serrated alternate ovate leaves. Flowers have 5 or more petals with spectacular colors ranging from white to red. After pollination, the fruits that set are capsules having several brown, black seeds.

Hot and cold beverages are made by steeping the flowers in cold water or by boiling the flowers. The "tea" thus prepared is garnished with lemon juice or other flavoring agents plus sugar or sugar substitutes and consumed as a beverage. The beverage is popular in Egypt West Africa Mexico Caribbean countries and in parts of Asia including India, Cambodia, and Vietnam. Studies conducted by USDA scientists show that Roselle tea *H. Subdariffa* consumption lowers blood pressure due to inhibition of Angiotensin.

Horse chestnut (*Aesculus hippocastanum*) Kingdom: Plantae: Group: Angiospermae, Dicotyledonae, order: Sapindales Family: Sapindaceae.

Horse chestnut is a deciduous tree growing to a height of 118 ft. with a domed crown. The leaves of the plant are palmate compound with 57 leaflets per leaf. Flowers are produced in panicles with about 25 flowers per panicle. One to five fruits develop per panicle. Seeds inside the spiky fruit are also known as conkers. Seeds have about 20% Aescin that is efficacious in treating chronic venous insufficiency varicose veins and hemorrhoids. Both whole seed extract and pure aescin have antiinflammatory properties and are effective in controlling chronic venous insufficiency disease to a level as good as compression stockings. However, the dosage must be critically evaluated since raw horse chestnut has Saponins and glycosides, which are toxic. The glucoside aesculin in unprocessed horse chestnut is a potent toxin and must be removed during processing. The FDA has not approved any product of this plant for medicinal purposes. **Usage:** Standardized horse chestnut powder having no more than 100-150 mg of aescin and a gel having 2% aesin are available commercially for treatment of swelling of legs due to chronic venous insufficiency, varicose veins and hemorrhoids.

Indian snakeroot (*Rauvolfia serpentina*), Kingdom: Plantae, Group: Angiospermae, Eudicots, Order: Gentianales, Family: Apocyanaceae, Chromosome number 2n=22.

Rauvolfia canescens/R. Tetraphylla also of the same family has similar alkaloids. This species however grows in the carribean islands, Mexico, Indonesia, and India. It has similar medicinal use as *R. serpentina*

Plants are shrubs growing in the sub Himalayan and hill regions of the Indian subcontinent. Roots and other plant parts produce several indole alkaloids of which reserpine is important because of its anti-hypertensive and anti-psychotic effects. The active compound reserpine was marketed under the name's reserpine, serpacil, Raudixin for controlling hypertension. However, it is no longer prescribed because of its side effects. However, herbal extracts of the plant are still prescribed by traditional herbalists for treating mental illness as well as high blood pressure.

Jiaogulan (*Gynostemma pentaphyllum*) Kingdom: Plantae; Group: Angiosperms, Eudicots, Order: Cucurbitales; Family: Cucurbitaceae; Subfamily: Nhandiroboideae syn. Zanonioideae.

Gynostemma pentaphyllum is a climbing vine belonging to the cucumber family. The plant is a native of southern China mother Vietnam South Korea and Japan. The plants are either male or female. Both males and females' plants must be grown together to achieve pollination and production of fruits. There are over 30 species related to Gynostemma. The plant has been used in Chinese traditional medicine for its adaptogenic and antioxidant properties that are believed to increase longevity.

Health relevance: Plant extracts are Adaptogens since they regulate blood pressure, lower cholesterol, increase immune activity, inhibit tumor formation, control diabetes as well as protecting the liver from liver diseases. Usage: Leaves are used to prepare herbal tea or leaf powders are used to prepare capsules or tablets.

Lily-of-the-valley (*Convallara majalis*): Kingdom: Plantae, Group: Angiospermae, Monocots, Order: Asperagales, Family: Asperagaceae,

Plants are herbs with underground rhizomes. The rhizome extracts have many phytochemicals including Convallatoxin. All parts of this plant are very poisonous because they have cardiac glycosides. It is considered by herbalists as an adjunctive or standalone therapy for the treatment of cardiac disorders such as arrhythmia, mitral valve prolapses and shortness of breath.

Lily-of-the-valley can cause side effects such as nausea, vomiting, abnormal heart rhythm, headache, decreased consciousness and responsiveness, and visual color disturbances.

Pomegranate (*Punica granatum*), Kingdom: Plantae, Eudicots, Lamiales, Lamiaceae. Order: Myrtales, Family: Lythraceae

Plants are shrubs or small trees growing to a height of about 16 ft producing globose fruits having many juicy pearls of seeds. The juicy fruit seeds as well as the fruit rind are rich in polyphenols, anthocyanins, flavanoids and other antioxidants. Like green tea, pomegranate juice has been consumed for centuries with the belief that the ruby red fruit promoted health. Modern scientists have shown that this belief is well justified. Powerful antioxidant chemicals in pomegranate fruit and juice may help reverse atherosclerosis and lower blood pressure. The fruit juice is also a good diuretic.

Tea (*Camellia/Thea sinensis*): Kingdom: Plantae, Group: Angiospermae, Eudicots, Order: Ericales, Family: Theaceae. Chromosome number 2n= 30.

The plants are bushes/shrubs growing in sub-tropical weather. The young leaves are preferred for beverages although older leaves are also used for making tea of a poorer quality. Many studies claim that drinking green Tea is good for sustain good health because of its antioxidant properties. Green tea is said to lower blood pressure and reduce stroke, lower cholesterol.

OTHER: The following plant products have also been reported to control high blood pressure (hypertension) and other cardiovascular problems: Buchu (*Agathosma betulina*-Family: Rutaceae); Celery (*Apium graveolens*-Apiaceae), Mu tong (*Aristolochia manshuriensis*-Family Aristolochiaceae), Oats (*Avena sativa*-Family: Poaceae/Gramineae), Coffee weed (*Cassia occidentalis*-Family: Caesalpiniaceae), Black bean/Moreton Chestnut

(*Castanospermum austral-* Family: Fabaceae), Karpurvali (*Coleus forskohlii/C.barbatus*-Family: Lamiaceae), Virginia mayflower (*Commelina virginica*)(Family: Commelinaceae), Chinese Hawthorn (*Crataegus pinnatifida* (Family: Rosaceae), River Lily, Swamp Lily (*Crinum glaucum*-Family: Amaryllidaceae), Hardy Fuchsia/Chiko/Tilco (*Fuchsia magellanica*-Family: Onagraceae), Flaxseed/, Linseed, (*Linum usitatissimum*: Family: Linaceae), Kizha Nelli/Nela nelli (*Phyllanthus amarus/P. niruri*-Family: Euphorbiaceae), Maritime Pine (*Pinus pinaster*-Family: Pinaceae), Sticky Nightshade/Wild Tomato (*Solanum sisymbriifolium*-(Family: Solanaceae), Common Wheat bran (*Triticum aestivum*-(Family: Poaceae/Gramineae), Cat's Claw herb (*Uncaria rhynchophylla*-(Family: Rubiaceae), Black plum (*Vitex doniana*-(Family: Verbenaceae).

References

Web and internet

https://www.texasheart.org/heart-health/heart-information-center/topics/digitalis-medicines.

https://www.britannica.com/science/digitalis.

http://www.clevelandheartlab.com/blog/top-herbs-for-your-heart.

https://www.pacificcollege.edu/news/blog/2014/07/14/herbs-heart-health.

https://www.acc.org/about-acc/press-releases/2017/06/12/14/57/traditional-chinese-medicine-may-benefit-some-heart-disease-patients.

https://www.webmd.com/heart-disease/news/20170612/meds-rooted-in-ancient-china-may-help-heart-review#1.

https://www.ncbi.nlm.nih.gov/pmc/articles/PMC4279313.

Books and Journals

Hollman, A (1985): "Plants and cardiac glycosides" (PDF), Br. Heart J. 54 (3): 258–261. doi: 10.1136/hrt.54.3.258. PMC 481893, PMID 4041297.

Jain, S, Buttar, H.S., Chintameneni, M., Kaur (2018): Prevention of Cardiovascular Diseases with Anti-Inflammatory and Antioxidant Nutraceuticals and Herbal Products: An Overview of Pre-Clinical and Clinical Studies. Recent Pat Inflamm Allergy Drug Discov, 2018; 12(2): 145-157. Doi: 10.2174/1872213X12666180815144803.

Karthikeyan K, Bai BR, Gauthaman K, Sathish KS, Devaraj SN. (2003): Cardio protective effect of the alcoholic extract of Terminalia arjuna bark in an in vivo model of myocardial ischemic reperfusion injury. Life Sci. 2003; 73: 2727–39.

Khare CP. (2007): New Delhi: Springer (India) Pvt. Ltd; 2007. Indian Medicinal Plants – An Illustrated Dictionary. 2nd Indian Reprint; p. 33.

Nadia Afsheen, Khalil-ur-Rehman, Nazish Jahan, Misbah Ijaz, Asad Manzoor, Khalid Mahmood Khan, and Saman Hina (2018): Cardio protective and Metabolomics Profiling of Selected Medicinal Plants against Oxidative Stress. Oxid Med Cell Longev, 2018; 2018: 9819360. Published online 2018 Jan 14, doi: 10.1155/2018/9819360, PMCID: PMC5821957, PMID: 29576858.

Nahida Tabassum and Feroz Ahmad (2011): Role of natural herbs in the treatment of hypertension (2011): Pharmacogn Rev. 2011 Jan-Jun; 5(9): 30–40. doi: 10.4103/0973-7847.79097, PMCID: PMC3210006, PMID: 22096316.

Nishteswar, K. (2014) Credential evidences of Ayurvedic cardio-vascular herbs. Ayu. 2014 Apr-Jun; 35(2): 111–112.doi: 10.4103/0974-8520.146194. PMCID: PMC4279313. PMID: 25558152.

Rastogi S1, Pandey MM2, Rawat AKS2 (2017): Spices: Therapeutic Potential in Cardiovascular Health, Curr Pharm Des. 2017, 23(7): 989-998. doi: 10.2174/1381612822666161021160009.

Rastogi, S, Pandey, M.M., Rawat, A.K., Traditional herbs (2016): a remedy for cardiovascular disorders. Phytomedicine, 2016 Oct 15, 23(11): 1082-9.

Stansbury, J; Saunders, P, Winston, D, and Zampieron, E. R. (2012): he Use of Convallaria and Crataegus in the Treatment of Cardiac Dysfunction. Journal of Restorative Medicine, Volume 1, Number 1, pp. 107-111.

CHAPTER 7

Herbals for Digestive System, Diabetes

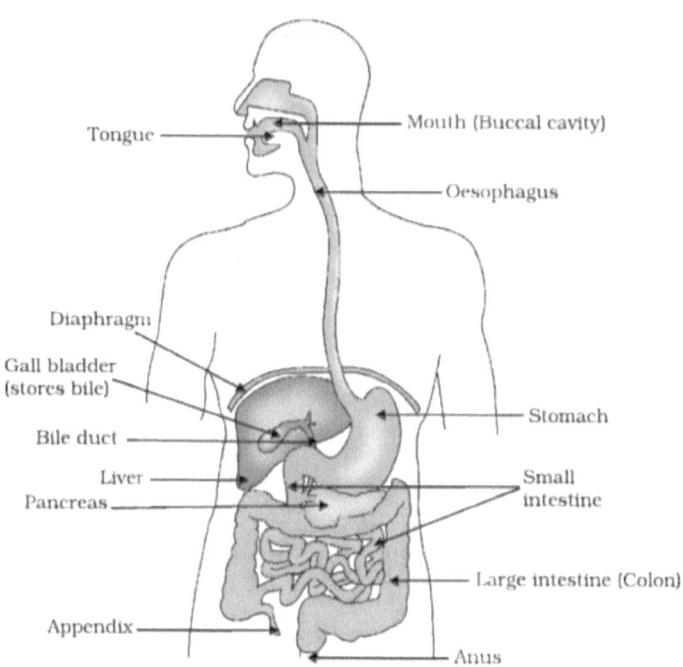

The digestive system consists of the mouth, esophagus, stomach, the small and large intestines, the anus, the pancreas, gallbladder, and liver. In the mouth, the food is chewed, and enzymes chiefly amylase is added by secretions from the salivary glands. From the mouth, the food moves down through the esophagus into the stomach wherein the enzyme pepsin and HCl secreted by gastric secretory cells are added to digest the food. Contractions of the stomach wall churn the partially digested enzyme acid mixture to form a semi solid mixture of partially digested food called chyme. Additionally, other secretory cells add mucosa to the food mash.

From the stomach, the partially digested food moves into the duodenum over a period of about four hours. The duodenum secretes a hormone, cholecystokinin (CCK), which causes the gall bladder to contract, releasing alkaline bile into the duodenum. CCK also causes the release of digestive enzymes from the pancreas. Enzymes from the pancreas mainly insulin, inactive trypsinogen which is a precursor of trypsin and lipase as well as bicarbonate are added in order to neutralize the acidic food and digest the protein, carbohydrates and fats in the food. Moreover, bile juice secreted by the liver and stored in the gall bladder is added to the food mash in the duodenum. The bile acts to emulsify fats present in the food thus helping digestion of fats by lipases. Large amounts of nutrients released by digestion in the duodenum are absorbed there. The rest of the digested food moves down into the rest of the small intestine where the nutrients from the digested food are absorbed. The undigested remnants along with water are pushed into the large intestine also known as colon by intestinal peristalsis. The water in the digested food is reabsorbed in the colon. If the reabsorption of water is excessive then the result is constipation. On the other hand, if reabsorption is poor, the result is diarrhea. The large intestine has many different bacteria that produce certain essential vitamins, which are absorbed by the intestine and are crucial to prevent vitamin deficiency. Hence, good colon health depends on the presence of probiotic micro flora. Long-term antibiotic therapy will destroy the microbial ecology of the colon and hence during antibiotic treatment it is beneficial to replace the microbial flora in the colon by taking probiotics. The completed digested food moves into the anal region from where the debris is evacuated.

Major diseases of the digestive system include cancers and ulcers of the mouth, salivary glands, esophagus, stomach, small and large intestines and the liver and pancreas. Gallstones in the gall bladder are also causes for gastric distress. Cancers of the stomach, duodenum, pancreas, liver, and colon are major diseases affecting the intestinal digestive system. Chronic acid reflux disease due to stomach acids refluxing back into the esophagus and mouth causing heartburns is another common ailment of the digestive system. The other major disease of the digestive system is diabetes due to improper production of insulin by the pancreas. Currently, research shows that foods rich in fiber and antioxidants coupled with exercise can cut down the incidence of cancers in the digestive system as well as control of diabetes. Foods that are rich in sugars and saturated fats should be avoided. Vegetables and fruits, which are rich in nutrients and antioxidants, are beneficial for digestive system health.

Appetite Stimulants

Drinking a small amount of a bitter drink before a meal is conducive to stimulating appetite. This usually takes the form of alcoholic drinks such as Chartreuse, Compari, Cyanar, Angostura and a host of others. However, traditional herbal concoctions include non-alcoholic drinks having different herbal extracts. Examples are Dandelion leaves (*Taraxacum officianale*), Gentian root (*Gentiana lutea*), Artichoke flowers (*Cyanara scolymus*), Indian Bitter melon fruits (*Momordica charantia*), and Broccoli florets (*Brassica oleracea*). In all cases, the herbs/vegetables are boiled in water and the boiled water extract is consumed in small amounts before a meal or the water extracts are mixed with alcohol.

Carminatives

Carminatives are herbs that reduce flatulence and are anti-gas products. In general, these are mostly spice seeds or extracts of spice seeds and are consumed after a meal. Examples are 1. Ajwain (*Trachyspermum ammi*), 2. Basil leaves/seeds (*Ocimum basilicum*) 3. Cardamom seeds (*Eletteria cardomom*), 4. Caraway seeds (*Carum carvi*), 5. Cumin seeds (*Cuminum siminum*) 6. Dill (*Anethum graveolens*), 7. Ginger rhizome (*Gingiber officianalis*).

Demulcents and Other Digestive Aids

Demulcents are herbs that sooth the passage of foods by inducing secretion of mucilage in the digestive tract. Other Digestive aids help in the digestive process through enzymes present in them or they prevent colic and irritable bowel activities. Important herbs in this group are 1. Slippery Elm bark (*Ulmus fulva*), 2. Milk Thistle (*Silybum marianum*), 3. Peppermint (*Mentha *piperata*), 4. Chamomile (*Matricaria recutita*), 5. Licorice (*Glycyrrhiza glabra*), 6. Lemonbalm, (*Melissa officinalis*), 7. Turmeric (*Curcuma longa*)

Bowel Movement

The gastrointestinal transit time (GTT) which is the time taken for food from entry into mouth through digestion and emptying out is 53 hrs. However, it is much less for people on a high fiber vegetarian diet. After the digestive process in the stomach, duodenum and small intestine, the undigested food along with the fibrous waste moves down into the large intestine aided by peristaltic movements. The probiotic microorganisms in the act on the food waste by fermentative processes as a source of microbial nutrition and in the process; they generate vital nutrients such as vitamin B-12 needed by the human body for hematopoiesis. This is of great importance because vitamin –B12 is absent in vegetarian and vegan diets. A proper balance of the micro biome of the colon is important for good digestive health; an imbalance can cause diahorrea or constipation, flatulence, colitis, and irritable bowel syndrome. Apart from natural inhabitants *Escherichia coli*, yeast, and other natural inhabitants, enriching the colon ecology by adding other probiotics can help proper digestion. *Lactobacillus rhamnosus*, *Lactobacillus acidophilus*, *Bifidobacterium lactis* and *Bifidobacterium longum* through yogurt as well as probiotic capsules will aid and help prevent colitis, IBS, and other intestinal inflammatory processes. They will also help in smooth movement of the waste food through the large intestine into the anus and eventual discharge.

One of the most common complaints that many people have is the hardening of the fecal matter and inability to discharge the waste resulting in constipation. A food rich in meat/protein without fiber leads to constipation. Both soluble fiber as well as insoluble fiber can draw water into the large intestine and thus smoothing the digested food waste and enabling movement and discharge. Foods rich in fiber such as whole grains, legumes, vegetables, and fruits should be included in everyday diets to improve colonic health and proper prompt bowel movements. In addition, the seed husk of certain herbs such as *Plantago ovata* (blond Psyllium/Isabgol/Indian Plantago), extract of leaves and pods of Senna

(*Senna alexandrines*), Triphala (mixture of three fruit extracts from Amla/ Nellikkai (*Emblica officinalis*), Bibhitaki (*Terminalia bellirica*), and Haritaki (*Terminalia chebula*)act as laxatives that help in the relief of constipation.

Liver

The liver is an accessary organ located in the right upper quadrant of the abdomen just below the diaphragm that is very much involved in the digestive process, In addition to detoxification, emulsification of fats through production of bile, various complex biological functions, the liver is also a store house of essential vitamin B complex, Iron and other nutrients. The liver takes part in at least 500 different metabolic activities in the body. Abnormalities of liver include hepatitis caused by Viruses Hepatitis A, B, C, D, E, amoebic hepatitis, liver cirrhosis, hepatomegaly, jaundice, fatty liver, liver cancers, hepatic encephalopathy, and certain genetic conditions. Alcohol and certain drugs also cause liver dysfunction. Thus, proper functioning of the liver is critical. Plants that are considered beneficial for liver function are 1. Milk thistle (*Silybum marianum*), 2. Long pepper (*Piper longum/Piperis Longi*), 3. Holy Basil/Thulasi (*Ocimum sanctum*), 4. Gale of the wind/Kizha Nelli (*Phyllanthus niruri*), 5. Herbal mix of *Capparis spinosa, Cichorium intybus,* Mandur bhasma (calcined iron oxide), Solanum *nigrum, Terminalia Arjuna, Cassia occidentalis, Achillea millefolium, Tamarix gallica* marketd as Liv 52.

Gall Bladder

The gall bladder is the storehouse of bile generated by the liver. The bile flows from the gall bladder into the duodenum through the bile duct. The bile emulsifies fats in the food. Besides gall stones which can block the gall tube other diseases include inflammation due to stones (Choleocystitis), inflammation without stones (Acalculous gallbladder disease), scarring of bladder and duct due to unknown causes (Sclerosing cholangitis), growths (polyps) and rare gall bladder cancer. The herbals that are used to treat gallstones are Apple cider, vinegar with apple juice, Dandelion, Milk thistle, Artichoke, Psyllium husk and Castor oil pack.

Pancreas

Pancreas plays an extremely important role in producing and secreting sodium bicarbonate into the acidic choam and amylase, lipase, trypsin respectively and thus aid digestion of carbohydrates, fats, and protein. The insulin secreted by the islet of Langerhans of the pancreatic region regulates and breaks down sugar. Thus, this endocrine organ has multiple roles. Since pancreatic enzymes are not stable in the acidic environment of the stomach, they must be enteric coated to help them function. Therefore, although several plant-derived enzymes are available, they must be packaged in enteric-coated tablets or pills for efficiency.

Biology of plants useful for good digestive health: The plants described here have multiple medicinal uses. For the sake of convenience, these plants have been listed under different sub-headings since these are the major uses for these plants.

CARMINATIVES (ANTI-FLATULENT), APPETITE STIMULANTS, AND OTHER DIGESTIVE AIDS

Ajwain/Ajwain/Carum/Bishops weed/Omum, (*Trachyspermum ammi/ Carum copticum/Ammi copticum*): Kingdom: Plantae; Division: Angiospermae, Class: Eudicots Order: Apiales Family: Apiaceae. Chromosome number, 2n=18.

Ajwan seeds are used as carminatives (anti-flatulent), appetite stimulant and digestive aid. The seeds can be distilled to produce an essential oil rich in terpenes. Seeds soaked in water or seed extracts known as gripe water is commonly used to treat colicky conditions. Ajwan plants are annual herbs, producing fruits that are botanically called schizocarps but commonly called seeds.

Angelica (*Angelica archangelica*): Kingdom: Plantae, Division: Eudicots, Order: Apiales, Family: Apiaceae.

The roots of the plant are used for medicinal purposes as an appetite stimulant, anti-flatulent and general digestive aid. It is also used for a multiple of other conditions such as urinary infections, fever, bronchitis, and colds. Root is used in preparation of absinthes, aquavits, and bitters-liquors. The roots and seeds **have coumarins** and hence it is not recommended for use by patients already on blood thinning agents. Plants are biennial herbs grown for its medicinal properties in the Scandinavian countries, Russia, France, and East European countries. The plant grows vegetatively in the first year and then it produces an elongated stem from which multiple florets arise in umbels.

Anise/Shombu/Anishuna: (*Pimpinella anisum*): Kingdom: Plantae, Division: Angiospermae, Class: Eudicots, Order: Apiales, Family: Apiaceae. Chromosome number, 2n=18 or 20.

Eating a few Seeds of Anise help in controlling intestinal spasms, bloating and colic. Anise seed extracts are components of certain cough lozenges, expectorants, and breath fresheners. Anise is used to flavor various liqueurs like the Greek **ouzo** and the French **anisette**. Large dosages have a narcotic effect and can also cause bleeding.

Plants are herbs native to Middle Eastern countries growing to a height of 3 ft or more. The plants have simple leaves at the base and at the top the leaves are feathery pinnate, divided into many leaves. The flowers are yellow or white 1/8 inch produced in dense umbels like those of carrots.

Artichoke flowers (*Cyanara scolymus*): Artichoke or Globe artichoke (*Cynara cardunculus var. scolymus*); Kingdom: Plantae: Division: Angiospermae, Class: Eudicots; Order: Asterales; Family; Asteraceae. Chromosome number: 2n=32

Artichoke flower heads reduces the risk of arteriosclerosis and reduces intensity of irritable bowel syndrome and helps overcome functional dyspepsia, which is a condition showing upper abdominal fullness, heartburn, nausea, belching, or upper abdominal pain. The Cynarin in the artichoke inhibits taste receptors and hence it makes even water to have sweet taste.

The edible flower head consists of many unopened flower heads in a cluster with many bracts. The plants are perennials growing to a height of 6 ft. with lobed large green leaves. The fleshy flower head is cooked by boiling in water.

Asafoetida/Peumkaayam/Hing (*Ferula asafoetida*): Kingdom: Plantae; Division: Angiospermae, Class: Eudicots Order: Apiales Family: Apiaceae; Genus: Ferula Species: F. asafoetida. Chromosome number, 2n=22.

In alternative medicine, it is used as an anti-flatulent digestive aid that eases colic and aids in treating irritable bowel syndrome. Asafoetida has Ferulic acid that is being considered as an anti-ageing chemical.

Ferula plant is a perennial herb growing in Afghanistan, Iran, Pakistan, and India. It is a monoecious plant with separate male and female flowers. Plants grow to a height of about 5 ft with hollow stems about 5 inches in diameter. The rhizome and taproot produce a white gummy resin that dries to a dark brown color. It has a fetid smell and the gum was known as devils' dung or food of the Gods or Hing in Hindi.

Basil/Holy Basil/Thulasi (*Ocimum tenuiflorum/O. sanctum*): Kingdom: Plantae; Division: Angiospermae, Class: Eudicots, Family: Lamiaceae/Labiatae; Chromosome number, 2n=32.

Holy Basil also known as Thulasi in India is an aromatic shrub with green (Lakshmi Thulasi) or purplish green leaves (Krishna Thulasi) and small racemate pink/white flowers. The entire plant is aromatic due to the presence of an essential oil. The Thai basil (*O. basilicum*), the hybrid Thai lemon basil and the hoary basil (*O. americanum*) are close relatives. The holy basil has been used for thousands of years in Ayurveda for its diverse healing properties. It is an adaptogenic and elixir of life. It is said to increase longevity and is used to treat multiple diseases including coughs and colds, respiratory illnesses, skin, and digestive diseases. Basil is used for stomach spasms, loss of appetite, intestinal gas, kidney conditions, fluid retention, head colds, warts, and worm infections.

Caraway/Kala Jeera/Perum Jeerakam (*Carum carvi*): Kingdom: Plantae; Division: Angiospermae; Class: Eudicots Order: Umbellales, Family: Apiaceae. Chromosome number, 2n = 20.

Caraway oil and seeds improve digestion and relieve spasms in the stomach and intestines. Traditionally, Caraway is used for digestive problems including heartburn, bloating, gas, loss of appetite, and mild spasms of the stomach and intestines. Caraway oil is also used to help people cough up phlegm, improve control of urination, kill bacteria in the body, and relieve constipation. Caraway seeds are used to flavor baked foods, liquors, rice dishes and in cosmetic preparations. Caraway plants are biennial with feathery leaves and small pink or white flowers in umbels.

Cardomom (*Eletteria cardomom*): Kingdom: Plantae, Group: Angiospermae, Class: Monocotyledonae Order: Zingiberales Zingiberaceae. Chromosome number Mysore varieties: 2n=48, 52, Malabar varieties 2n=48, 50.

Elettaria cardamomum is a perennial bushy plant native to tropic regions, and can grow to heights of 10' or more, with the tall stems showing long, alternate leaves.

Cardomom seeds **have terpineol, limonene, menthone, cineol, myrcene**, and other phytochemicals and are chewed as breath fresheners, stimulants of saliva and added to formulations to relieve pulmonary congestion.

Dill/Shatapushpaa (*Anethum graveolens*): Kingdom: Plantae; Division: Angiospermae, Class: Eudicots Order: Apiales; Family: Apiaceae. Chromosome number, 2n=20.

Dill seeds are anti-flatulent carminatives and help digestion. Seeds and oil have anti-microbial properties. Plants are herbaceous, growing to a height of 2 ft with slender stems and feather-like narrow leaves. Flowers grow on umbels. Seeds have an aromatic essential oil.

Chirayita/Kirata/Indian Gentian (*Swertia chirayita*): Kingdom: Plantae, Division: Angiospermae, Class: Eudicots, Order: Gentianales Family: Gentianaceae, Chromosome number. Synonyms are *Gentiana chirayita*, *Ophelia chirata*.

Swertia chirayita is an erect hardy annual growing to 1 m (3ft 3in) by 0.5 m (1ft 8in) growing in Himalayan foothills of India and Bhutan. Plants have lanceolate leaves and bear hermaphrodite flowers in large panicles. Chirata has a yellow bitter acid, **Ophelic acid**, two bitter glucosides, **Chiratin** and **amarogentin** and many other phytochemicals. The ethanol extract of *S. chirayita* exhibits hypoglycemic activity. The roots and plant extracts are components of the Ayurvedic preparation known as Mahasudarshna Churna. Chirata is used for fever, constipation, upset stomach, and loss of appetite, intestinal worms, skin diseases, and cancer. The plant extracts are used in various liquor bitters as appetite stimulants.

Cumin/Jeerakam/Jeerak (*Cuminum cyminum*): Kingdom: Plantae; Division: Angiospermae, Class: Eudicots Order: Apiales Family: Apiaceae. Chromosome number, 2n=14.

Cumin is a culinary and medicinal herb seed that is related to parsley. Decoctions, infusions, tablets, and seeds are used as digestive aids to help digestion and reduce flatulence. The seeds are also components of various Ayurvedic and Siddha formulations. Seed extracts have anti-microbial properties.

Plants are annual herbs with branched stems and pinnate or bi-pinnate leaves. Flowers are borne in umbels. Fruits are achenes with a single seed. Seeds are ridged and similar in appearance to caraway seeds. Dried seeds are used as such or after powdering .**Fennel/ Mishreya (*Foeniculum vulgare*):** Kingdom:

Plantae; Division: Angiosperms; Class: Eudicots, Order: Apiales Family: Apiaceae, Umbelliferae Genus: Foeniculum, Species: F. vulgare. Syn: *Officianalis, F. dulce*. Chromosome number, 2n=22.

Seeds are known carminatives and digestive stimulants. It is a part of many cough syrups, juices, lozenges to control cough, bronchitis, and asthma.

It is a hardy perennial herb with a bulbous stem, yellow flowers and feathery leaves and are found naturally growing in the Mediterranean countries but now naturalized in many parts of the world. The plants are highly aromatic, and the oil of fennel is a part of the popular liqueur absinthe. Plants grow to a height of about 10 ft. Leaves are divided into threadlike blades. Flowers are tiny, produced in umbels. Seeds are very tiny about 10 mm long. About 5% of the dried seed is made of an essential oil that has the stilbene foeniculoside. The seeds are used as a flavoring agent in foods and culinary purposes.

Ginger/Inji/Aadraka (*Zingiber officinale*): Kingdom: Plantae Division: Angiospermae; Class: Monocots, Order: Zingiberales, Family: Zingiberaceae. Chromosome number, 2n=22. Many polyploid varieties exist.

Ginger rhizome is used in traditional medicine to stimulate saliva (sialagogue) and thus digestion as well as to control nausea and vomiting. The rhizome has an essential oil namely, **gingerol**, which is used as a flavoring agent. Ginger has many other uses in herbal medicine.

Licorice/Yashtimadhu (*Glycyrrhiza glabra*): Kingdom: Plantae, Division: Eudicots, Order, Fabales, Family: Fabaceae

Plants are perennial herbs growing in India, Asia, Africa, and Southern Europe belonging to the legume family. Plants grow to a height of 3 ft with pinnate leaves and flowers in inflorescences. The sweet flavoring agent Liquorice is isolated from the stolon-like roots. The roots have **glycyrrhizin**, which is sweet. Liquorice is a prescribed medicine as per the Chinese traditional medicine and Indian Ayurvedic systems for many ailments particularly digestive conditions such as stomach ulcers, heartburn, reflux, colic, and ongoing inflammation of the lining of the stomach (chronic gastritis). However, Liquorice should be used with caution since clinical evidence indicates that it can raise blood pressure, edema and weight gain and worsen kidney health.

Malabar tamarind/Kudam Puli/Vrkshaamla (*Garcinia gummi gutta* syn *G. gambojia*): Kingdom: Plantae, Class: Angiospermae, Eudicots, Order: Malphigiales/Guttiferales, Family: Guttiferae

Malabar tamarind are large evergreen trees growing in Indonesia, Malay peninsula, Philippines, Southern China and in the slopes of the western Ghats of Southern India, as well as certain Carribean islands and Florida in the USA. The Plants grow to a height of over 80 ft and produce global fruits with red thick hard exocarp (Skin) and juicy citrus-like edible endocarp. The entire fruit is processed as treatment for various ailments including digestive conditions.

These sparsely branched trees are now grown all over the world. The large, serrated leaves are borne on the crown of the plant and the trunk has a scarred appearance due to the location of the early leaves, which have fallen off. Plants may be entirely male (male flowers only) or female (Female flowers) or bisexual with stamens and pistil in same flower. Female plants produce only small fruits and produce few fruits since they must be cross-pollinated. Growers plant only plants that are hermaphrodite (bisexual flowers). The entire plant is Laciferous. Fruits are oblong or elongated and when ripe have several black seeds. Both the unripe and ripe fruits as well as latex have proteolytic digestive enzymes. Because of its proteolytic properties, the fruit as well as latex and plant extracts are marketed as digestive supplements. *Garcinia Cambogia* has an ingredient called **hydroxycitric acid (HCA)**, which has been used to aid weight loss. The extract of HCA is available in powdered form or pill form and can be bought online or in health stores. However, clinical studies do not fully substantiate the weight loss claims made by many marketing agencies. The results have been mixed with some studies showing little or no weight loss effects while others claim substantial decrease in body weight. The Hydroxy citric acid in these fruits lowers the clotting of blood and hence should be used with caution.

Mangosteen/Purple mangosteen (*Garcinia mangostana*) Kingdom: Plantae, Class: Angiospermae, Eudicots, Order: Malphigiales/ Guttiferales, Family: Guttiferae

Mangosteen fruits with a purple rind or exocarp are grown on evergreen trees in South East Asia including Malaysia, Indonesia, Cambodia, Vietnam, Caribbean islands, Puertorico, Columbia, and other subtropical/tropical regions. The purplish edible fruits have white juicy sweet orange-lozenge like vesicles. The purple rind has xanthonoids. in addition, rinds are used to treat gastrointestinal diseases in South East Asia. People in these countries have used the pericarp (peel, rind, hull or ripe) of Garcinia as a traditional medicine for the treatment of abdominal pain, diarrhea, dysentery, infected wound, suppuration, and chronic ulcer.

Marshmallow (*Althaea officinalis*) Kingdom: Plantae, Division: Eudicots, Order: Malvales, Family: Malvaceae.

Plants are related to Hibiscus. They are shrubs growing to a height of 3-6 ft with glabrous ovate 5-lobed leaves and pinkish flowers. The leaves, roots and flowers are used in traditional medicine. The roots have the spongy marshmallow that is a good demulcent and soothing agent for treating gastritis and is used in culinary dishes. However, the commercial marshmallows no longer have the extract from this plant.

Peppermint (*Mentha* × *piperata*): Kingdom: Plantae, Division: Angiospermae, Class: Eudicots, Order: Lamiales, Family: Lamiaceae.

It is a well-known digestive herb. Peppermint can be taken in hot infusions or used to make a peppermint tea. It is available as enteric-coated capsules for the relief of IBS, available in health stores. Enteric-coated peppermint powder and oil capsules prevent dyspepsia. Peppermint oil is also used in confectionaries. The plant is a hybrid cross of spearmint and water mint. It has growth characters like mint.

Slippery Elm Bark (*Ulmus rubra/U. fulva*): Kingdom: Plantae, Angiospermae, Eudicots, Rosales, Ulmaceae

The bark powder used as a tea decoction or sprinkled on breakfast cereals is a digestive herb remedy for digestive discomfort, such as acid dyspepsia, irritable bowel syndrome, stomach cramps and hyperacidity. It is approved as an over the counter demulcent.

Plants are deciduous trees, which are closely related to American elm and other elm species. The bark of the trees holds a mucilaginous carbohydrate.

Spearmint (*Mentha spicata*): Kingdom: Plantae, Division: Angiospermae, Class: Eudicots, Order: Lamiales Family: Lamiaceae, Chromosome number, 2n=48.

Mint leaves are chewed to help in digestive processes. Mint may cause allergic reactions in some people. Plants are perennials that grow in both temperate and tropical climates. Leaves are serrated and range in colors from deep green to light blue and purple depending on the variety. The plants can spread fast and may be invasive.

Triphala (mixture of three fruit extracts from Amala/Nellikkai (Emblica officinalis), Bibhitaki (*Terminalia bellirica*), and Haritaki (*Terminalia chebula*): Kingdom: Plantae, Division: Eudicots. Order and family T. *bellerica* and T. *Chebula*: Myrtales, Family: Combretaceae, Order and Family for *Emblica officianalis/Phyllanthis emblica*: Order: Malphigiales, Family: Phyllanthaceae.

The fruit extracts from all three plants are combined into a common herbal medicine called Triphala (meaning three fruits). Triphala is used as a carminative, mild gentle laxative and is also recommended as a treatment for diabetes. Both species of Terminalia are deciduous trees growing in the Indian sub-continent as well as other countries of south east Asia. The fruits are dried and powdered and are two of the constituents of Triphala. *Terminalia chebula* is also an excellent laxative.

Phyllanthus emblica is also a tree with compound leaves. The fruits known as amla are dried and used as a third part of Triphala. The fresh fruit is used in pickles and fruit extracts as well as leaf extracts are components of hair oil tonics.

Diabetes

Bitter melon/Bitter gourd/Karela (*Momordica charantia*): Kingdom: Plantae, Class: Angiospermae, Eudicots, Order: Cucurbitales, Family: Cucurbitaceae

Plants are climbing vines with simple wavy-margin leaves. Flowers are yellow in color and are unisexual. Fruits are elongated gourds with many hard-shelled seeds. Fruits as well as leaves are bitter. Fruits are used in culinary dishes, particularly in the Indian subcontinent. Chinese and introduced American varieties are longer, plumper and less bitter while the Indian varieties are smaller, spinier, and very bitter. The fruits and fruit extracts are popular supplements because of their hypoglycemic effects. An **insulin analog polypeptide –p** has been isolated from this plant. Some studies show that the capsule and tablet supplements are less effective as hypoglycemic agents compared to fruit extracts (drinks). More studies are needed.

Black Plum/Java Plum/Jambool/Naval pazham (*Syzygium cumini* also known as *Eugenia Jambolana* or *E. Cumini*): Kingdom: Plantae, Class: Angiospermae, Eudicots, Order: Myrtales, Family: Myrtaceae.

The plants are tall trees native to Indian subcontinent and Australasia. It is now grown in Florida, USA as well as the Caribbean islands. The bark, fruits and seeds have multipurpose medicinal values. It has good anti-diabetic and cardio protective properties. It is an also considered good for protecting the liver. The fruits are rich in polyphenols, anthocyanins and flavanoids.

Blond Psyllium/Isabgol/Indian Plantago (*Plantago ovata*): Kingdom: Plantae, Division: Eudicots, Order: Lamiales, Family: Plantaginaceae.

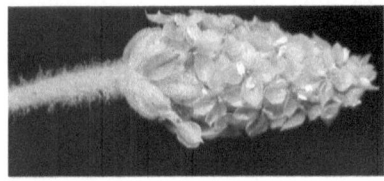

Plantago plants are herbs growing to a height of about a foot and are grown commercially in India, Pakistan, Soviet Union, and many European countries. It grows wild in the Southwestern United States. Plants have long narrow leaves with an elongated flower stalk from which spikelet of flowers arise. The plants are grown for extracting the soluble fiber from the husk of seeds. The husk fiber is an excellent treatment for regularizing bowel movements and treating constipation. The soluble fiber from the husk is a part of many commercial products for helping smooth bowel movements. It is also an approved treatment choice for reducing cholesterol uptake since the fiber binds cholesterol. It is also approved by FDA for reducing blood sugar in treating diabetes.

Cloves/Lavanga (*Syzygium aromaticum*): Kingdom: Plantae, Class: Angiospermae, Eudicots, Order: Myrtales, Family: Myrtaceae

Cloves are the flower buds of the evergreen tree that is native to the Moluccas islands of Indonesia and grown extensively in the state of Kerala in India. The cloves are used to flavor foods and are considered to have anti-diabetic properties. The USDA has not approved use of cloves or clove oil for treatment of diabetes. The flower buds are good mouth fresheners and clove oil numbs mouth gums and so is a good anesthetic for control of toothache.

Fenugreek (*Trigonella foenum-graecum*): Kingdom: Plantae, Division: Angiospermae, Class: Eudicots, Order: Fagales, Family: Fagaceae/ Leguminosae. Genus and species: Trigonella foenum-graecum. Chromosome number, 2n=16.

Fenugreek is used for treating digestive problems such as loss of appetite, upset stomach, constipation, and inflammation of the stomach (gastritis). Whole and powdered seeds and leaves slow absorption of sugars in the stomach and stimulate insulin. The seeds have **saponins, soluble fiber, trigonelline alkaloid and coumarin.** The seeds and seed extract lower blood sugar in people with diabetes. Both fresh and dried leaves are used as vegetables and the yellow seeds are used as a spice in the Indian subcontinent. Because of Coumarin content, consumption of fenugreek by people on blood thinners is not recommended.

Fenugreek is an annual herbaceous plant with leaves consisting of three small obovate to oblong leaflets. It is cultivated worldwide as a semiarid crop and its seeds are a common ingredient in dishes from the Indian Subcontinent.

Indian Gooseberry/Amla (*Embelica officianalis*): Indian gooseberry (*Emblica officinalis/Phyllanthus emblica*): Kingdom: Plantae; Division: Angiospermae, Class: Eudicots, Order: Malphigiales; Family: Chromosome number= 2n=14, Polyploids varying from 98-104,

All parts of the plant but especially the fruits are prescribed for various illnesses by practioners of both Indian traditional Ayurvedic, Unani and Siddha medicine as well as Chinese traditional medicine. The fruits are components of the ancient Ayurvedic rejuvenative medicine known as Chavanaprash that is considered good for digestion and bowel health. Some studies show that the fruits or fruit extracts reduced pancreatitis in rats and helped recovery of the pancreas after pancreatitis attacks. There are also reports that the fruits can help control of diabetes and improve diabetic neuropathy. The fruits are also antioxidants.

The Amla tree is a deciduous tree with simple pinnate leaves. The edible fruit is greenish-yellow with six vertical stripes wherein each striped area stands for a segment of the fleshy fruit. The seed inside the fruit has a hard shell and is brownish in color. The fruit upon wounding quickly turns brown to black due to polyphenol oxidase activity.

Insulin Plant/Fiery Costus/Spiral Flag (*Chamaecostus cuspidatus*): Kingdom: Plantae, Angiospermae, Monocots, Order: Zingiberales, Family: Costaceae

This fleshy herbaceous plant is cultivated in India because of its sugar lowering properties. It grows in India, Brazil, Florida, and many south American and Caribbean regions. Plants produce fleshy long leaves with parallel veins like ginger and turmeric. They produce attractive flowers. The plants also have underground ginger-like rhizomes. The dried leaf powders are prescribed by Ayurvedic and Siddha medicine practioners for treating type 2 diabetes.

Jack Fruit/Chakka/Jaca dura/Jaca mole/Jaca manteiga (Artocarpus heterophylla/A.integrifolia/A, braziliensis): Kingdom: Plantae, Angiosperms, Eudicots, Order: Rosales, Family: Moraceae.

The plants are large trees with thick hard wood stems. The plants grow in the Indian subcontinent, Sri Lanka, Thailand, Malayan Archipelago, Philippines, Guam and in south and Central America including the Carribean islands. The plants produce separate male and female flowers in an inflorescence in the same tree. After pollination, all the flowers get enclosed in a compound fruit called the Jack fruit. Each individual fruit within the compound fruit has a hardy thick oval shaped seed which is rich in protein and complex carbohydrates. The sweetish edible portion encloses the seeds. The edible fruit contains natural soluble and insoluble fiber as well as fruit sugars. The fruit is said to improve insulin sensitivity. The seeds can be toasted or powdered after they are sun dried. Both the fruit and seed meal help control diabetes. As per the International Glycemic index service operated by the University of Sydney, raw but edible jack fruit has a low glycemic index.

Salacia/Chundan/, Kothala Himbutu/, Ponkoranti (*Salacia reticulata/S. oblonga*): Kingdom: Plantae, Angiosperms, Eudicots, Order: Celastrales, Family: Celastraceae

Plants are perennial climbing shrubs native to Sri Lanka and Indian subcontinent. The plants are important medicinal herbals as per Ayurveda. The bark and leaf decoctions drunk as tea controls diabetes and lowers the glycated form of hemoglobin (Hemoglobin A 1 c). It is also recommended as an herbal for reducing weight. The plant extracts are also prescribed for treating itchy skin and for inflamed joints.

Sugar destroyer/Charkkarai Kolli/Gurmar/Madhunashini (*Gymnema sylvestre*): Kingdom: Plantae, Class: Angiospermae, Eudicots, Order: Gentianales, Family: Apocyanaceae

Plants are climbers with soft green leaves. Leaves and plant extracts have **Gymnemic acid** which interacts with taste receptors in the mouth and thereby controlling the taste of sugar. Extracts of the plant consumed as capsules, tablets, lozenges, and juices control the taste of sugar and reduce the craving for sugary food. Plant extracts consumed along with anti-diabetic medicines have been shown to control both Diabetes 1 and 2.

Malabar Plum/Java Plum/Jamun/Naval pazham (*Euginea jambolena/ Syzygium cumini*): Kingdom: Plantae, Class: Angiospermae, Eudicots, Order: Myrtales, Family: Myrtaceae

The plants are native to the Indian sub-continent, Australasia and are introduced species in Caribbean islands, South America, and Florida. Fruits are ovoid global with a purple skin and juicy flesh. The fruits and juice are helpful in controlling diabetes. Both the seeds and the fruit are diuretics and have important carminative and astringent properties. The seed extracts and fruit juices reduce blood sugar levels and are useful in the treatment of diabetes. The seeds and bark are well known in the Far East for the treatment of dysentery and in the control of hyperglycemia and glycosuria in diabetic patients. The fruit extracts are also prescribed for control of liver diseases.

Potential Anti Cancer and Anti Inflammation Plants of Digestive System

Broccoli: (*Brassica oleracea*): Kingdom: Plantae: Division: Angiospermae, Class: Eudicots; Order: Brassicales; Family; Cruciferae, Chromosome number: 2n=18.

Recent studies at the University of Pittsburgh have shown that broccoli sprout extracts protect against development of mouth cancers and the researchers have coined the term "Green chemoprevention" for treatment protocols based on broccoli.

Broccoli is a member of the italic group of *Brassica oleraceae* related to cabbage, kale, Brussels sprouts, Mustard, and bok choy. The plant can grow to height of 1 ft-8 ft. high. The flower head is the vegetable.

Turmeric/Haridra (*Curcuma longa*): Kingdom: Plantae, Class: Angiospermae, Monocots, Order: Zingiberales, Family: Zingiberaceae

Turmeric plants are perennial herbs with a yellow-saffron rhizome that has finger-like projections. The above ground leaves are long and tapered with parallel veins. The rhizomes are usually sun-dried and powdered. The powder is a major part of curry powders and Indian as well as south Asian cuisine. The rhizomes have curcumin that is reported to have anti-cancer and anti-polyp properties in the intestinal tract. Curcumin is poorly absorbed into the blood stream but is better absorbed if black pepper extracts are added or if the curcumin is encapsulated in lysosomes. More information is provided in other Chapters also.

Stomach Ulcers, Hemarrhoids

Calendula/marigold (*Calendula officinalis*): Kingdom: Plantae, Angiospermae, Division: Eudicots, Order: Asterales, Family: asteraceae.

The plants are perennial herbs growing mostly in southern Europe and in warmer temperate countries. It is also grown as an annual in colder climes. The plants have spectacular chrysanthemum-like yellow or orange color flowers. The flowers petals

have flavanoids, **lutein, zeaxanthin and terp**inoids. The petal extracts are used for treatment of stomach ulcers, rectal inflammation (proctitis) and hemorrhoids.

Black nightshade/Manathakkali/Black nightshade (Solanum nigrum) Kingdom: Plantae, Angiospermae, Division: Eudicots, Order: Solanales, Family: Solanaceae.

The plants are herbs which grow wildly in Australasia, Caribbean, Southern United States, South America. Plants produce tiny pea size green fruits which become black upon maturity. The leaves are heart shaped with serrated edges. Both leaves and fruits are edible. The unripe fruits are toxic, but the friend ones have less of the toxic solanine. It has been described as a remedy for digestive issues in ancient Greek medicine, Indian Ayurveda, and Siddha medicine and in other folk medicines. It is used after cooking for treating mouth ulcers, herpes, general digestive issues.

Liver Health

Kutki/Indian Gentian (*Picrorhiza kurroa*): Kingdom: Plantae, Angiospermae, Eudicots, Order: Lamiales, Family: Plantaginaceae.

Kutki is an herbaceous plant growing in the Himalayan ranges. The rhizome of the plant has hepato- protective chemicals and hence is a popular and often prescribed medicinal herb for treating liver cirrhosis and liver inflammation.

Kizha Nelli/Tamalaki/Stonebreaker (*Phyllanthus niruri*/*P. amarus*): Kingdom: Plantae, Angiospermae, Eudicots, Malphigiales, Phyllanthaceae

These plants are small (2 ft tall) herbaceous plants found in tropical and semi tropical coasts. Plants have tiny green compound leaves with small capsular fruits with seeds that grow on the underside of leaves. The whole

plant and fruits are considered to protect the liver from hepatitis diseases and dissolve stones as per Ayurvedic protocols. However, the Cochrane Database of Systematic Reviews does not verify anti-viral claims.

Long pepper/pippili (*Fructus piperis longi/Piper longum*): Plantae, Angiospermae, Magnoliids, Piperales, Piperaceae.

Extracts of the fruits of this plant are considered to have hepato (Liver) protective abilities. As well as digestive aids. A study has shown that the extract of *Fructus Piperis Longi* treats liver diseases by reducing the activities of transglutaminase, such as serum aspartate amino transaminase, alanine aminotransferase, alkaline phosphatase, and gamma-glutamyltransferase, which are the main causes of the development of liver cirrhosis, and by reducing bilirubin (total, direct and indirect) content, which leads to jaundice. The ethanol extract of long pepper was found to have superior activity against hepatitis B virus *in vitro*. In rodents, this plant was assessed for its hepatoprotective effect against carbon tetra chloride -induced acute, chronic, reversible, and irreversible liver damage.

Pippali is a slender aromatic climber with leaves, which are entire, glabrous, with reticulate venation. The creamy-colored flowers are borne in solitary cylindrical spikes, the male flowers longer and more slender than the female spikes, the latter giving way to a cylindrical cluster of small ovoid fruits about 4 cm in length, that passes from green to orange-red in color when ripe, becoming black upon drying. Pippali is found growing wild throughout the hotter regions of Southeast Asia in evergreen forests but is cultivated extensively in the state of Kerala in India for Ayurvedic medicinal preparations.

Milk Thistle (*Silybum marianum*): Plantae, Angiospermae, Eudicots, Order: Asterales and Family: Asteraceae

Milk thistle extracts and powders, which contain the active principle **silymarin**, are being used and prescribed by alternative medicine as a

 treatment for control of liver diseases including jaundice due to Hepatitis A, C as well as liver cirrhosis and cancer. But various controlled tests reviewed by the Cochrane report or the United Kingdom (UK) cancer review do not verify these claims, but Milk thistle products continue to be marketed in Health food stores.

Milkweed plants are small thorny herbs growing wild in most temperate countries. The plants have oblong pinnate leaves with spiny edges. The stem and leaves exude a milky exudate and hence the name milk thistle. Leaves and plants are toxic to animals. The marketed supplements often have mycotoxins and hence care should be taken to ensure that the products are free from toxins.

Papaya (*Carica papaya*): Kingdom: Plantae, Class: Angiospermae, Eudicots, Order: Caricales, Family: Caricaceae

Papaya plants are tall unbranched trees with large serrated long petiolalated leaves. The lower leaves fall off and leave scar marks. The Stem leaves and all other parts of plant release latex upon injury. The latex is corrosive and has many digestive enzymes. The pants can be male, female, or bisexual. The fruits can be globose or elongated depending on variety. There are two basic types of plants those producing yellow colored fruits and those with red fleshy pulp. The leaves are used in herbal medicine for treating malaria and for inducing abortions. A decoction of leaves is also prescribed for diabetic control. Fruits are good for helping digestion.

Laxatives and Anti-Constipation Agents

Cascara/bearberry (*Cascara sagrada/Rhamnus purshiana*): Kingdom: Plantae, Division: Eudicots, Order: Rosales, Family: Rhamnaceae.

Cascara is a North American native shrub plant whose bark have **Anthraquinones**. Anthraquinones have laxative properties. Intake of Cascara bark extracts cause peristaltic movements pushing the stools down into the colon where it swells and causes pressure to evacuate. It will induce bowel movement in about 12 hrs. after intake. The drug was approved by FDA initially, but FDA approval was rescinded in 2002 due to safety concerns. Cascara is still available as a supplement in health food stores.

Castor beans (*Ricinus communis*) Kingdom: Plantae, Class: angiospermae, Eudicots, Order: Malphigiales Family: Euphorbiaceae. Castor bean plants are shrubs mostly growing in tropical and subtropical areas. The oil from seeds known as Castor oil is a powerful laxative (Purgative). It is also a vermifuge. The leaf extracts have hepato protective activities as shown in animal experiments. All plants of the plant especially the seed kernels have the cytotoxic ricin.

Senna/Tinnevelly Senna/Egyptian Senna/East Indian Senna (*Senna alexandrina*), Kingdom: Plantae, Division: Eudicots, Order: Fabales, Family: Fabiaceae.

Plants are shrubs growing to a height of about 4 ft. The plants bear leathery leaflets on opposite leaves. Flowers are yellow borne on racemes. Pods bear about 4-6 seeds. The Senna glycosides are laxatives and hence seed extracts are used as laxatives both in allopathic as well as traditional medicine. Senna is one of the essential medicines in the World Health Organization (WHO) list of essential medicines. Senna usage should be only for limited periods since long-term usage does not result in laxative action. Further, the use of Senna is not allowed in cases of abdominal obstruction, colitis, pain, and other abdominal inflammatory conditions

Senna-aavaram/Aavarai (*Cassia auriculata*): Kingdom: Plantae, Division: Eudicots, Order: Fabales, Family: Fabiaceae.

Aavaram Senna is a close relative of Tinnevelly Senna and the plants have similar morphology. The leaves have laxative properties and the roots are anti-diabetic. Flowers and flower buds are powdered to make a tea that is anti-diabetic.

Tamarind (*Tamarindus indica; T. occidentalis T. officinalis*): Kingdom: Plantae, Class: Angiospermae, Eudicots, Order: Fabales, Family: Fabaceae

Tamarind trees are huge trees growing to a height of over 60 ft with very thick serrated/pitted stems. They are natives of tropical Africa and have been introduced to South Asia particularly India which is the largest producer of Tamarind. The plants grow well in South Florida (USA) and in Mexico and Caribbean islands. Trees bear compound pinnately lobed leaves. The fruits are pods with a brown pulp covering brown seeds. The fruit pulp is very sour and tangy having tartaric, malic, and other organic acids along with B vitamins. The fruit is a mild laxative and helps relieve constipation. The dry seeds are rich in protein and can be pan fried, powdered, and consumed.

Wild Castor/Danti (*Baliospermum montanum/Jatropha montana,*), Kingdom: Plantae, Class: Angiospermae, Eudicots, Order: Euphorbiales, Family: Euphorbiaceae

Plants are shrubs growing in the Sub Himalayan and peninsular regions of India, Plants have simple glabrous slightly serrated leaves. Flowers are unisexual produced in racemes. Root extracts and paste are applied to shrink hemorrhoids and to ameliorate abdominal pain. Seed extracts are laxatives and anti-helminthic.

The following plants described elsewhere are also considered as herbal aids for digestive conditions; Bilberry, whortleberry: (*Vaccinium myrtillus*) – lower cholesterol, lower blood sugar; Cinnamon Ceylon (*Cinnamon Cinnamomum verum/C. Zeylanicum*) - Lower sugar levels; Garlic (*Allium sativum*) – lower cholesterol and blood sugar can cause stomach irritation. Guduchi/Amrit/Giloy (*Tinospora cordifolia*) - general health and digestive aid; Malabathrum (*Cinnamomum tamala*) is another plant whose leaves and bark have sugar lowering abilities.

Probiotics

The gastrointestinal system is populated by many microorganisms. They play important roles in generating digestive enzymes and nutrients like vitamin B-12. They also keep in check potentially harmful pathogens from gaining control and causing disease. Extensive use of antibiotics will kill many beneficial bacteria and allow yeast-like Candida to grow and cause thrush, candidiasis, vaginal infections, pruritus, and diarrhea. So, use of probiotics, vegetables, and fruits or at least during antibiotic therapy help in restoring the ecological balance in the digestive tract. Regular consumption of yogurt also aids the balancing of gut biome. *Lactobacillus rhamnosus, Lactobacillus acidophilus, Bifidobacterium lactis* and *Bifidobacterium longum* are some useful probiotic organisms.

References

Web and Internet

https://www.sciencedirect.com/journal/food-research-international/vol/44/issue/7.

https://www.ncbi.nlm.nih.gov/pmc/articles/PMC5442075.

https://www.mskcc.org/cancer-care/integrative-medicine/herbs/tribulus-terrestris.

https://www.mskcc.org.

https://www.ncbi.nlm.nih.gov/pubmed/29617079.

https://en.wikipedia.org/wiki/Trees_of_India.

Books and Journals

Ajay C. Donepudi, Lauren M. Aleksunes, Maureen V. Driscoll, Navindra P. Seeram, and Angela L. Slitt (2012): The traditional Ayuverdic medicine, Eugenia Jambolana (Jamun Fruit) decreases liver inflammation, injury, and fibrosis during cholestasis. Liver Int. 2012 Apr; 32(4): 560–573. PMCID: PMC3299847, NIHMSID: NIHMS340494PMID: 22212619

Akhila J Shetty, Divya Choudhury, Rejeesh, Vinod Nair, Maria Kuruvilla, and Shashidhar Kotian (2010): Effect of the insulin plant (Costus igneus) leaves on dexamethasone-induced hyperglycemia, Int J Ayurveda Res. 2010 Apr-Jun; 1(2): 100–102, doi: 10.4103/0974-7788.64396, PMCID: PMC2924971, PMID: 20814523.

Baby Joseph and D Jini (2013): Antidiabetic effects of Momordica charantia (bitter melon) and its medicinal potency, Asian Pac J Trop Dis. 2013 Apr; 3(2): 93–102.; doi: 10.1016/S2222-1808(13)60052-3; PMCID: PMC4027280.

Ca Gianluca Ianiro, Silvia Pecere, Valentina Giorgio, Antonio Gasbarrini, and Giovanni Cammarota (2016), Digestive Enzyme Supplementation in Gastrointestinal Diseases, Curr Drug Metab. 2016 Feb; 17(2): 187–193.Published online 2016 Feb. doi: [10.2174/138920021702160114150137], PMCID: PMC4923703, PMID: 26806042ldecott, Todd (2006). Ayurveda: The Divine Science of Life. Elsevier/Mosby. ISBN 0-7234-3410-7.

Fei Xiong and Yong-Song Guan (2017): Cautiously using natural medicine to treat liver problems, World J Gastroenterol 2017 May 21, 23(19): 3388–3395, Published online 2017 May 21. doi: [10.3748/wjg. v23.i19.3388], PMCID: PMC5442075, PMID: 28596675

Joseph B, Jini D. Insight into the hypoglycaemic effect of traditional Indian herbs used in the treatment of diabetes Res J Med Plant 2011a; 5(4): 352–376.

Khanna P, Jain SC, Panagariya A, Dixit VP. (1981): Hypoglycemic activity of polypeptide-p from a plant so Asian Pac J Trop Dis. 2013 Apr 3(2): 93–102.

Pucci ND, Marchini GS, Mazzucchi E, Reis ST, Srougi M, Evazian D, Nahas WC (2018) Effect of Phyllanthus niruri on metabolic parameters of patients with kidney stone: a perspective for disease prevention. Int Braz J Urol. 2018 Jul-Aug, 44(4): 758-764. Doi: 10.1590/S1677-5538.IBJU.2017.0521.

Roxas M. The role of enzyme supplementation in digestive disorders. Altern. Med. Rev. 2008; 13(4): 307–.

Vishal, R. (2013): Protective role of Indian medicinal plants against liver, The Journal of Psychopharmacology 2013; 2(3): 1-3, on line, http://www.phytopharmajournal.com.

Zakaria, S. Moreover, Hanan El-Kabany (2009) Effects of Fructus Piperis Longi extract on fibrotic liver of gamma-irradiated rats, Chin Med. 2009; 4: 2. PMCID: PMC2657146, PMID: 19183455. Published online 2009 Jan 30. doi: [10.1186/1749-8546-4-2].

CHAPTER 8

Genito-Urinary System and Plants

The urinary system consists of the two kidneys, the ureters, bladder, and urethra. The genital system, which is closely associated with the urinary system, consists mainly of the prostate along with ejaculatory glands, the scrotum, testes, penis, epididymis, Vas deferens and urethra in males and in females the ovaries along with fallopian tubes, the uterus, cervix, vagina, and clitoris. The adrenal glands that produce cortisol are attached to the kidneys. The urinary and the genital systems are associated.

The kidneys take part in filtering and removing body waste, regulate blood volume and pressure, electrolyte balance and pH. The urinary system is highly vasculated. After filtration, the waste in the form of urine is pumped into the bladder through ureters and from the bladder, it is expelled through urination. Both female and male urinary tracts are similar in anatomy, physiology, and function.

Diseases of the urinary system include Kidney stones, infection, and resultant burning -feeling, incontinence, kidney failure, cancers, interstitial cystitis, overactive bladder, kidney cysts and other kidney related problems due to diabetes.

As for the male reproductive system, the main medical conditions are prostatitis, male erectile dysfunction, low sperm count and non-optimal ejaculatory conditions. In the case of females, most conditions relate to menopausal symptoms such as hot flashes, weight gain, osteoporosis, dryness of vagina, cancer of ovary and vagina and cysts.

The following herbs are sources and components of traditional herbal formulas for treatment of various genitourinary conditions. These herbs have been listed in specific groups of health conditions for convenience although they may have multiple effects and fall into multiple categories.

Abortifacients

Bitter gourd (*Momordica charantia*): Kingdom: Plantae, Clade: Angiosperms, Eudicots, Order: Cucurbitales, Family: Lamiaceae Family: Cucurbitaceae

Botanical description has been given earlier Chapter 7. The fruits are used as a vegetable and as a drink for diabetic control. It is also an abortifacient.

Oleander (*Nerium indicum; N. odorum, Thevetia peruviana*), Kingdom: Plantae, Clade: Angiosperms, Eudicots, Order Gentianales, Family: Apocyanaceae.

Leaves, flowers, and fruits especially seeds have many toxic cardiac glycosides. Seed extracts are used as abortifacient.

Ochrosia (*Ochrosia oppositifolia*), Kingdom: Plantae, Clade: Angiosperms, Eudicots, Order Gentianales, Family: Apocyanacease. This African plant is a local carminative in the Seychelles (Africa) but in large quantities, it is used as an abortifacient.

Rosemary (*Rosmarinus officinalis*): Kingdom: Plantae, Clade: Angiosperms, Eudicots, Order Lamiales, Family: Lamiaceae. Although rosemary leaves and oil are used as fragrance and in culinary preparations, it is also considered to have unsubstantiated abortifacient properties.

Snake gourd (*Trichosanthes cucumerina* and *T. kirilowii*): Kingdom: Plantae, Clade: Angiosperms, Eudicots, Order Cucurbitales, Family: Cucurbitaceae.

The snake-like fruits of this plant are popular vegetables in South Asia and China. The fruits of some varieties can be very bitter. The fruits have abortifacient properties due to presence of the phytochemical **Trichosanthin**.

Other Abortifacients Reported in the Literature

Black musli/Kali musli/Nilappana: (*Curculigo orchioides*): Kingdom: Plantae, Clade: Angiosperms, Monocots, Order Asperagales, Family: Hypoxidaceae: Paste of the tuber is given orally in empty stomach to induce abortions. It is also a medicine for treating erectile dysfunction and some urinary and digestive conditions (described elsewhere).

Chaff-flower/prickly chaff flower/devil's horsewhip/Apaamaarga-Sanskrit: अपामार्गः (*Achyranthus aspera* Lin) Kingdom: Plantae, Clade: Angiosperms, Eudicots, Order Caryophyllales, Family: Amaranthaceae: Fresh root is made into paste and it is mixed with lukewarm water and given as an abortifacient.

Doctor *Bush/Ceylon leadwort/Wild leadwort/, Koduveli=Tamil-India (Plumbago zeylanica* L): Kingdom: Plantae, Clade: Angiosperms, Eudicots, Order Caryophyllales, Family: Plumbaginaceae: 3-5 ml of root paste is administered orally to induce abortion.

Garudakkodi/Eswaramooli/swarmul (*Aristolochia indica* Linn) Kingdom: Plantae, Clade: Angiosperms, Magnolids, Order Piperales, Family: Aristolochiaaceae: Fresh root is used as an abortifacient. Seed paste is given orally on empty stomach for 3-5 days. 3-5 ml of bark juice is taken orally for the first three months for inducing abortions.

Papaya/Kappalam (*Carica papaya* L.): Kingdom: Plantae, Clade: Angiosperms, Eudicots, Order Caricales, s, Family: Caricaceae: 10-15 ml of latex of raw fruit is given orally once a day for 3 days.

Pineapple/Annasi/(*Ananas comosus* L.): Kingdom: Plantae, Clade: Angiosperms, Monocots Order: Commelinids, Family: Bromeliaceae. Fruits have **B gallic acid, syringic acid, vanillin, ferulic acid, sinapic acid, coumaric acid, chlorogenic acid, epicatechin, and arbutin.** Ripened Pineapple fruit is used to induce abortion. It is contraindicated for anyone on Coumadin (blood thinners) treatment.

Poinciana/peacock flower,/red bird of paradise/Mexican bird of paradise/dwarf Poinciana/pride of Barbados,/flos pavonis, and flamboyant-de-jardin/Rajamally(-Tamil-India: (*Caesalpinia pulcherrima* (L.): Kingdom: Plantae, Clade: Angiosperms, Eudicots, Order Fabales, Family: Fabaceae: Bark juice (2 ml) is administrated orally on empty stomach for the first three months of pregnancy.

Ramachempu (*Rhynchosia rufescens*): Kingdom: Plantae, Clade: Angiosperms, Eudicots, Order Fabales, Family: Fabaceae: Leaf decoction is administered for abortion for the first three months of pregnancy.

Turkey berry/prickly nightshade/shoo-shoo bush/wild eggplant/pea eggplant/pea aubergine/Chundakkai/Chunda/Ana chunda: (*Solanum torvum*). Kingdom: Plantae, Clade: Angiosperms, Eudicots, Order Solanales, Family: Solanaceae: 3-5 ml Leaf extract is given orally for 5days to induce abortions.

Aphrodisiacs, Erectile Dysfunction and Testosterone

Abarema (*Abarema sp.*): Kingdom: Plantae, Clade: Angiosperms, Eudicots, Order Fabales, Family: Fabaceae. These are trees growing in the Amazonian area under neo-tropical conditions. The tree bark extracts are used as aphrodisiac.

African cherry orange (*Citropsis articulata*): Plantae, Clade: Angiosperms, Eudicots, Order: Sapindales, Family: Rutaceae.

These are shrubby trees growing in Africa and are related to Citrus. In Uganda, an infusion made of the ground root, is considered a powerful aphrodisiac. Hence, the plants are threatened for extinction and are listed under conservation.

Catuaba (*Erythroxylum catuaba/vaccinifolium*): Kingdom Plantae, Group: Angiosperms, Division: Eudicots, Order: Malphigiales, Family: Erythroxylaceae. Known as caramuru, it is a Brazilian aphrodisiac with properties for treating erectile dysfunction.

Catharanthus (*Catharanthus trichophyllus*), Kingdom: Plantae, Group: Angiosperms, Division: Eudicots, Order: Gentianales, Family: Apocyanaceae. The plants are close relatives of the Madagascar periwinkle. Different varieties have different flower colors. The leaf extracts are an aphrodisiac and hemostatic. It is used to treat Sexually transmitted diseases and impotency.

Cow-Itch Plant/Cowhage/Velvet bean (*Mucuna pruriens/Mucuna prurita*), Kingdom: Plantae, Group: Angiosperms, Division: Eudicots, Order: Fabales, Family: Fabaceae.

Plants are annual climbing shrubs. The leaves, pods and all parts of the plant generate contact dermatitis (itchiness) due to presence of the protease Mucunain. However, the itchiness is removed upon cooking. Leaves are trifoliate as in beans, Flowers are white to lavender colored. Both Unani medicine and Ayurveda recommends the seeds as a treatment for increasing sperm count and management of male sexual dysfunction by increasing sperm count and testosterone levels. However, more commonly, seed pastes are used to treat snakebites. The seeds have high levels of L. DOPA and hence considered as a treatment choice for symptoms of Parkinson's disease.

Fenugreek (*Trigonella foenum grecum*): Kingdom: Plantae, Group: Angiosperms, Division: Eudicots, Order: Fabales, Family: Fabaceae/Leguminosae.

The leaves and seeds of this legume are considered good for control of Diabetes-2 and is an aphrodisiac. Seeds and leaves have coumarin (Warfarin) like compounds and hence can affect the anti-clotting effects of coumarin prescriptions.

Goat Weed (*Epimedium* sp.): Kingdom Plantae, Group: Angiosperms, Division: Eudicots, Order: Ranunculales, Family: Berberidaceae.

This is a perennial herb used in Chinese medicine to treat impotence and boost testosterone levels. Goat weed also increases nitrous oxide levels in the body, increasing blood flow to the sexual organs.

Hygrophila/Kokilaksha/Swamp weed (*Hygrophila* sp.) Kingdom: Plantae, Group: Angiosperms, Division: Eudicots, Order: Lamiales, Family: Acanthaceae.

Plant extracts are beneficial in treating impotence, spermatorrhea and seminal debilities.

India drumstick (*Moringa Indigofera*): Kingdom: Plantae, Group: Angiosperms, Division: Eudicots, Order: Brassicales, Family: Moringaceae.

The fruits and leaves are rich in vitamin C, flavanoids and is recommended by Ayurvedic practitioners for increasing libido/virility and sexual activities.

Kale/Leaf cabbage (*Brassica oleracea*): Kingdom: Plantae, Group: Angiosperms, Division: Eudicots, Order: Cruciferales, Family: Cruciferaceae.

This plant has indole-3-carbinol, a phytochemical that reduces the activity of the aromatase enzyme in the body. The aromatase enzyme takes part in the production of estrogen, the female hormone that blocks testosterone production. Reducing estrogen levels allows testosterone levels to increase.

Nilapana/Ground Palm (*Curculigo orchioides*): Kingdom: Plantae, Group: Angiosperms, Division: Eudicots, Order: Asperagales, Family: Hypoxidaceae (Amaryl-lidaceae).

This is a perennial herb partially described earlier also commonly grown in China, South Asia and in Southern India. The juice from rhizome is used for treating Erectile dysfunction, increase sperm count and as a medicine for good urinary function. It is also prescribed by Chinese and Kampo and Ayurvedic herbalists in formulations for digestive conditions for hepato protection.

Pepitas/Pumpkin seeds (*Cucurbita pepo*): Kingdom: Plantae, Group: Angiosperms, Division: Eudicots, Order: Cucurbitales, Family: Cucurbitaceae.

The seeds are nutrient rich having 30 % protein, monounsaturated fat 15%, polyunsaturated fat 20%, substantial levels of vitamins E and folate, as well as high levels of magnesium. The seeds per se are believed to boost testosterone levels.

Small Caltrops/Tribulus/devils weed/Devils thorn./Gneriginil/ Gokshura (*Tribulus terrestris*): Kingdom: Plantae, Group: Angiosperms, Division: Eudicots, Order: Zygophyllales, Family: Zygophyllaceae.

Tribulus plants are prostrate perennial creepers mostly found as invasive species in lawns, in wasteland and as weeds in Indian subcontinent and now in many countries including the North American continent. It is rarely cultivated except as a source for alternative medicinal use. Plants produce yellow colored bisexual flowers with five sepals, petals, stamens, and carpels. Fruits are green and mature to release five spiny nutlets. The fruits and nutlets are used in traditional medicine for chest pain, heart problems, dizziness, skin, and eye disorders, to expel kidney stones, and as a diuretic and tonic. *Tribulus terrestris* has potent diuretic properties, which increase blood flow to the kidneys. This process increases the amount of water excreted through urination, which is beneficial in combating cystitis and urethritis (swelling and irritation of the urethra). It is marketed as a dietary supplement to improve sexual function and for bodybuilding due to the belief that it acts like testosterone in the body. However, this effect has not been confirmed. Many studies also report liver toxicity in animals and two alkaloids present in this plant (harmane and nor harmane) cause staggers in sheep that graze and eat this plant.

Yohimbe (*Pausinystalia johimbe*): Kingdom: Plantae, Division Angiospermae, Eudicots, Order: Gentianales, Family: Rubiaceae

Yohimbe are trees that grow to a height of about 90-100 ft tall in many African countries particularly west and central Africa. The bark from these trees are stripped and extracted for preparations of herbal medicinal concoctions especially as aphrodisiacs. Although widely used to treat erectile dysfunction (LED), current guidelines do not recommend its use for treating erectile dysfunction. There is no scientific basis to support claims for ED or as aphrodisiacs.

Anti-Aphrodisiac

Chaste tree (*Vitex agnus-castus*): Kingdom: Plantae, Group: Angiosperms, Division: Eudicots, Order: Lamiales, Family: Lamiaceae

This plant also known as Monk's Pepper is an anti-aphrodisiac. Monks used to chew the berries and leaves of this plant to reduce sexual urges.

Diuretics, Prostate and Urinary Tract Issues

Cranberry (*Vaccinium oxycoccus/V. macrocarpon*), Kingdom: Plantae, Group: Angiosperms, Division: Eudicots, Order: Ericales, Family: Ericaceae

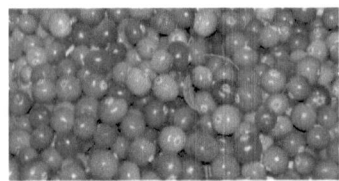

Cranberry plants are creeping perennials with four-lobed pink flowers and red global shiny berries growing in most countries in the Northern hemisphere. Cranberries are grown for making jams, sauces, and various culinary foods. Nevertheless, the berries also have potential medicinal values as treatment for urinary infections. Cranberries have compounds known as proanthocyanidins that have been shown to prevent bacteria from attaching to the bladder wall as also prevent bacteria from attaching to the stomach lining and areas in the mouth. Cranberry extracts are believed to control urinary tract infections (UTI), *Helicobacter pylori* infection and atherosclerosis. In patients who are prone to kidney stones, regular use of cranberry should be limited as it has oxalates that combine with calcium to cause stones in kidney. Animal and cell culture studies show that cranberry juice extract and isolated compounds can inhibit various types of cancer cells, but this has not been confirmed by clinical trials.

Dandelion (*Taraxacum officinalis*): Kingdom: Plantae, Angiospermae, Division: Eudicots, Order: Family, Compositae, Genus and species, Other common names: Blow ball, Puffball, pee-a-bed, wet-a-bed.

Dandelion plants are regarded as weed plants growing mostly in backyards, lawns, and roadsides but are now also grown as a vegetable crop for use of leaves and rhizomes in salads, smoothies, pesto, wraps and juice as additives. Plants are herbs belonging to the sunflower family with very bright yellow flowers and an underground stem -rhizome. The leaves on rhizomes have bitter phytochemicals that stimulate the bile and intestinal digestive enzymes. The rhizomes function as diuretics and hence the names pee-a-bed, wet-a-bed.

Indian Sarsaparilla/Sariva/Anantamul/Nannari/Maahali (*Hemidesmus indicus*): Kingdom: Plantae, Division: Angiospermae, Eudicots, Order: Gentianales, Family: Apocyanaceae.

The roots of this plant have a reno-protective effect, which protects against chemically induced nephrotoxicity (kidney toxicity). They are also mild diuretics. The roots are cut and used as pickles and are also added to sugary sherbets.

Matcha Green Tea (*Camellia sinensis*): Kingdom: Plantae, Group: Angiosperms, Division: Eudicots, Order: Ericales, Family: Theaceae.

Matcha is a specially grown and processed green tea powder popular in Japan and China. There are three grades of Matcha namely, ceremonial grade, premium grade, and culinary grade. These tea bushes grown in shade have more caffeine and theine than the darker varieties. They have high levels of antioxidants. Apart from Genito-urinary benefits, the green tea has good cardio-protective properties.

Saw Palmetto (*Serenoa repens*): Kingdom: Plantae, Group Angiosperms, Division: Monocot, Order: Arecales, Family: Arecaceae

Plants are monocotyledonous fan palms with about 20 leaflets per leaf ending in a fan with tapered ends. Each leaf including leaflets is about

1-2 meters long (3-6 ft). The flowers are held in compound panicles and fruits are drupes. Extracts from the fruits are claimed to control benign prostate hyperplasia (BPH) but these claims are not backed by the American cancer Institute or the National Center for Complementary and Integrative Health (NCCIH) of USA. There are also reports that suggest that extracts of Saw Palmetto interfere with the functioning of blood thinning agents. Nevertheless, saw palmetto extracts are compounded into various preparations marketed by the supplements industry as a panacea control of BPH.

Small Caltrops/Tribulus/devils weed/Devils thorn./Gneriginil/Gokshura (Tribu-lus terrestris): Described earlier, extracts of this plant have good diuretic properties.

Dissolving Kidney Stones

Black Cumin/Black Caraway/Kalonji/Karum Jeerakam/Kala Jeera (*Nigella sativa*), Kingdom: Plantae, Clade: angiospermae, Eudicots, Order: Ranunculales, Family: Ranunculaceae

They are annual herbs. Seeds are used in culinary preparations as well as in traditional medicine of Iran (Persia) for control of Kidney stones, dyspnea (shortness of breath), lowering blood pressure and cholesterol.

Caspian Manna/Camel thorn (*Alhagi maurorum*): Kingdom: Plantae, Clade: angiospermae, Eudicots, Order: Fabales, Family: Fabaceae

This is a short tree with extensive underground rhizomes. It is grown in Temperate Europe, Asia, and middle east as well as Afghanistan, Pakistan, and Northern India. The rhizome and leaf decoctions and powders are prescribed by Iranian herbologists in the province of Shiraz for dissolving Kidney stones as well as biliary conditions including jaundice.

Chanca Piedra/Kidney stone tree' Kizha Nelli (*Phyllanthus niruri, P. quebra*). Kingdom: Plantae, Division Angiospermae, Eudicots, Order: Malphigiales, Family Phyllanthaceae.

Leaves and the small fruits of this plant are prescribed by Ayurveda, Unani, and Siddha forms of medicine for liver and kidney ailments and dissolving kidney stone.

Coconut (*Cocos nucifera*): Kingdom: Plantae, Division Angiospermae, Monocots, Order: Arecales, Family Arecaceae

Pastes/extracts made of the flower of coconut in milk removes Kidney stones. Likewise, coconut water acts as a gentle diuretic and electrolyte drink

Copaibo/Palo de aceite (*Copaifera officinalis*): Kingdom: Plantae, Division Angiospermae, Eudicots, Order: Fabales, Family Fabaceae

The Tree is well known to the inhabitants of the Amazonian region for its medicinal properties. An essential oil prepared from the flowers and fruits of the pant are used as a treatment for colic, rheumatism and kidney cleansing including dissolving stones.

Daeba/Dharbha or Kusa grass/Halfa grass, Salt reed grass (*Desmostachya bipinnata*, Syn: *Eragrostis cynosuroides*): Kingdom: Plantae, Division Angiospermae, Monocots, Order: Poales, Family Poaceae.

Plants are grasses growing in India, N. Africa, Algeria, Ethiopia, Israel, and many other Middle Eastern countries in semi-arid regions. The leaves are used in religious ceremonies by Hindus and Buddhists. There are reports that show that the grass absorbs radioactive elements. Leaf extracts are prescribed as a diuretic to disgorge kidney stones and relieve pain.

Punarnavaa (*Boerhavia diffusa*): Kingdom: Plantae, Division Angiospermae, Eudicots, Order: Caryophyllales, Family Nyctaginaceae.

This plant native to the Indian sub-continent is considered a rejuvenating plant and apart from being prescribed for treating Kidney stones, formulations containing it are prescribed for treatment of joint inflammation, eye diseases, and skin conditions and as a general tonic.

Formation of Kidney Stones

The following plant products have oxalates or are high in phosphates that in turn can cause kidney stones to develop. At the same, time these plants have medicinally valuable chemicals, nutrients especially carbohydrates and protein. The oxalates can be removed/dissolved after or during cooking. These are 1. Celery (*Apium graveolens*). 2. Cranberry (*Vaccinium macrocarpon*), 3. Elephant yam (*Amorphophallus sp*) 4. Chinese potato/ Taro/Chembu/Cheppam kizhangu/Arvi (*Colacasiaesculenta and other species*), 5. Chocolate Cocoa (*Theobroma cacao*), 6. Spinach (*Spinacia oleracea*), 7. Tea (*Thea sinensis*), 8. Rhubarb (*Rheum hybridus*), 9. Cashewnuts/Kāju (*Anacardium occidentale*), 10. Peanuts (*Arachis hypogaea*), 11. Soybeans, (*Glycine max*) 12. Beets (*Beta vulgaris*).

Menstrual Issues

Abuta/Midwifes herb (*Cissampelos pareira*), Kingdom: Plantae, Group: Angiosperms, Division: Eudicots, Order: Ranunculales, Family: Menispermaceae.

Plants are climbing vines with heart-shaped leaves. Plants bear separate male and female flowers. Fruits are red drupes/berries. Extracts from all parts of the plant are used for treating pre-menstrual and menstrual problems.

Alligator Yam, Milky yam/Vidari kanda (*Ipomoea digitata/I. paniculata/I. mauritiana*), Kingdom: Plantae, Group: Angiosperms, Division: Eudicots, Order: Solanales, Family: Convolvulaceae.

Plants are perennial vines with light pink/violet flowers like those of morning glory and sweet potato. The rhizomes of the plants have **beta sitosterol** and **Ergonovine** (alkaloid). The rhizomes are used to stop menstrual bleeding. In India, Alligator Yam is used as a general tonic, to treat diseases of the spleen and liver and prevent fat accumulation in the body. The herb is also known as a lacto-stimulant and libido enhancer.

Asparagus/Shatavari (*Asparagus racemosa*): Kingdom: Plantae, Division Angiospermae, Monocots, Order: Asperagales, Family: Asperagaceae

The stem and reduced leaves that together are used in culinary dishes and salads. It has estrogenic properties, which alleviate symptoms of menopause, including anxiety, depression, mood fluctuations, insomnia, weight gain, irritability, and loss of bladder control.

Black cohosh/black bugbane/black snakeroot/fairy candle (*Actaea racemosa*), Kingdom: Plantae, Group: Angiosperms, Division: Eudicots, Order: Ranunculales, Family: Ranunculaceae

Plants are natives of the North American continent. They are herbaceous perennials. Flowers are borne on long racemic stalks/Flowers have no petals or sepals. Black cohosh is used in capsules as a dietary supplement for treating pre-menstrual tension and menopause symptoms. Evidence for these claims is sketchy.

Dong Quai: Known as 'Female Ginseng" (*Angelica sinensis*): Kingdom: Plantae, Division Angiospermae, Eudicots, Order: Apiales, Family: Apiaceae. This fragrant herb is used to treat period pain, amenorrhea (absent periods), Pre-Menstrual Syndrome (PMS) and symptoms of peri and menopause, such as hot flashes and lack of libido. However, the plants have potentially cancer-causing chemicals.

Licorice/Yashtimadhu (*Glycyrrhiza rubra*) Kingdom: Plantae, Division Angiospermae, Eudicots, Order: Fabales, Family: Fabaceae

Soothes the mucous membrane of the genital organs and alleviates perimenopausal symptoms including itchiness in the vulva, hot flashes, and night sweats

Linum/Flax (*Linum usitassimum*), Kingdom Plantae, Group: Angiosperms, Division: Eudicots, Order: Malphigiales, Family: Linaceae.

Flax plants are perennial plants grown for the fiber (linen and seeds for linseed oil. Plants grow to a height of 4 ft; have lanceolate shiny leaves with bluish flowers. Flax seeds are golden yellow or brown in color. Seeds have

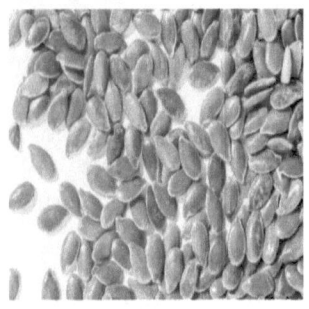
an oil rich in Omega-3 fatty acids and have a mucilaginous edible protein and a concentrated source of phytoestrogens, which have estrogen-like hormonal properties. Flaxseed may help improve menopausal symptoms and reduce blood pressure, but evidence of its ability to reduce cholesterol levels is mixed. These phytoestrogens are the reason flaxseed can affect menstrual cycle length and menopausal symptoms. Some studies showed flaxseed to be as effective as hormone replacement therapy in the management of menopausal symptoms. Studies of postmenopausal women showed that flaxseed supplementation improved the ratio of hormones that are thought to help prevent breast cancer. Studies in animals have shown promising results, but human data are lacking. Supplementation with Flax seed oil may help reduce triglycerides, reduce cholesterol, and potentially reduce diabetic symptoms.

Maca/Peruvian Ginseng (*Lepidium meyenii*): Kingdom: Plantae, Division Angiospermae, Eudicots, Order: Brassicales, Family: Brassicaceae.

Maca plants are annual herbs with an underground radish-like rhizomatous taproot from which rosette eaves arise. Maca root powder extracts have been used traditionally to improve stamina and sexual functioning. Studies have shown that Maca may help to increase libido and improve athletic performance. It may also affect some hormone levels in the blood. The evidence to support the use of Maca in improving sexual function is limited. A few clinical trials showed increase in sexual desire, sperm count, and motility. It may also improve sexual dysfunction caused by antidepressant use or menopause. Further study is warranted.

Nut Grass/Musta/Muthanga/Nagarmotha (*Cyperus rotundus* Linn.) Kingdom: Plantae, Division Angiospermae, monocots, Order: Cyperales, Family: Cyperaceae.

Plants are perennial sedges that produce long leafy triangular stems from an aromatic bulbous underground stem. Leaves are long, slender and have a prominent mid rib. Flowers are in inflorescences with individual flowers having three sepals, petals, stamens, and carpels. The rhizome and to a much lesser extent leaves have essential oils, flavonoids, terpenoids, and sesquiterpene. Nut Grass has a long history of medicinal use in the Ayurvedic, Siddha and Unani systems as well as Arabic, Greek and Chinese systems of traditional medicine. Its benefits have been documented in the Charaka Samhita, one of Ayurveda's prime texts. Traditional Chinese Medicine (TCM) credits nut grass with the ability to restore 'qi', the natural patterns in which the body functions. Although used to treat digestive problems, the underground rhizome is used in infusions to treat menstrual pain. It has various pharmacological activities besides the above.

Red Clover (*Trifolium pretense*): Kingdom: Plantae, Division Angiospermae, Eudicots, Order: Fabales, Family: Leguminosae This is rich in phytoestrogens called isoflavones, which helps to combat the intensity and frequency of hot flashes.

Soybean (*Glycine max*): Kingdom: Plantae, Division Angiospermae, Eudicots, Order: Fabales, Family: Leguminosae,

Soybean plants are bushy vines with trifoliate leaves. Plants produce seeds in pods. The seeds are rich in isoflavones. Intake of Soybean milk, tofu and other products reduce menopausal symptoms such as hot flashes, vaginal dryness, sleep problems and mood swings. In addition, soy products (protein) reduce LDL (bad) cholesterol levels and may increase HDL (good) cholesterol levels. There are also reports suggesting that high intake of soybean products reduce the risk of breast cancer.

References

http://ijpsr.com/

https://www.ncbi.nlm.nih.gov/pubmed/29617079

http://www.ville-ge.ch/musinfo/bd/cjb/africa/index.php?langue=an

https://www.zimbabweflora.co.zw/speciesdata/species.php?species_id=145190

http://www.ville-ge.ch/musinfo/bd/cjb/africa/recherche.php

http://naturaldatabase.therapeuticresearch.com/nd/PrintVersion.aspx?id=789&AspxAutoDetectCookieSupport=1

https://www.ncbi.nlm.nih.gov/pmc/articles/PMC5039998/

https://www.flowersofindia.net/catalog/slides/Common%20Nut%20Sedge.html

http://tropical.theferns.info/viewtropical.php?id=Pleioceras+barteri

http://tropical.theferns.info/

https://www.feedipedia.org/

https://www.analogforestry.org/resources/database

http://www.plantsoftheworldonline.org

https://www.ipni.org/n/81155-1

https://www.alwaysayurveda.com/

Bahmani, M, Baharvand-Ahmadi, B., Tajeddini, P., 3 Rafieian-Kopaei, M, and Nasrollah Naghdi (2016) Identification of medicinal plants for the treatment of kidney and urinary stones, J Renal Inj Prev. 2016; 5(3): 129–133, Published online 2016 Jul 27. doi: 10.15171/jrip.2016.27, PMCID: PMC5039998, PMID: 27689108.

Burrows, J.E., Burrows, S.M., Lötter, M.C. & Schmidt, E. (2018) Trees and Shrubs Mozambique Publishing Print Matters (Pty), Cape Town Page 810 (Includes a picture).

Lebrun, J.-P. Moreover, Stork, A. L. (1991-2015). Enumération desplantes à fleurs d'Afrique tropicale et Tropical African Flowering Plants: Ecology and Distribution, vol. 1-7. Conservatoire et Jardin botaniques de la Ville de Genève.

Nie Y, Dong, X, He Y, Yuan, T, Han, T, Rahman, K, Qin, and, Zhang Q. (2013): Medicinal plants of genus Curculigo: traditional uses and a phytochemical and ethno pharmacological review, J. Ethnopharmacol., Jun 3,147(3): 547-63. doi: 10.1016/j.jep.2013.03.066. Epub 2013 Apr 3.

Pirzada, A. M, Ali, H.H. Naeem, M., Latif, M., Bukhari, A.H., and Tanveer, A (2015): Cyperus rotundus L.: Traditional uses, phytochemistry, and pharmacological activities. Journal of Ethnopharmacology [20 Aug 2015, 174: 540-560].

Pucci, N.D, Marchini, G.S, Mazzucchi, E, Reis, S.T, Srougi, M., Evazian, D and Nahas W.C (2018) Effect of phyllanthus niruri on metabolic parameters of patients with kidney stone: a perspective for disease prevention, Int Braz J Urol. 2018 Jul-Aug, 44(4): 758-764. doi: 10.1590/S1677-5538.IBJU.2017.0521.

Rajesh Thachayil Philip, Krishnaraj M.V, Madhaiyan Prabu, R. Kumuthakalavalli (2012): Herbal Abortifacients used by Mannan tribes of Kerala, India International Journal of PharmTech Research 4(3): 1015-1017 July 2012.

CHAPTER 9

Herbals for Dermal (Skin) Health

The dermal system is the largest organ tissue system of the human body. It consists of three main layers namely the epidermis, dermis, and hypodermis. The three layers are held together by connective tissue. Blood vessels supply all three layers and the outer epidermal layer is covered by hair and sweat glands. Melanocytes found in the epidermis define the color of skin. Mutations in the melanin biochemical pathway can result in albinism, vitiligo (leucoderma), cancers and other skin pigment conditions. The black/brown melanin protects cells from harmful UV irradiation. The skin protects all the internal tissues and organs from microorganisms and environment, keeps body temperature, and reacts to touch and senses heat and cold.

The skin is subject to diseases due to both environmental causes as well as genetic reasons. The main types of skin conditions are 1. Rash, 2. Dermatitis (Eczema), 3. autoimmune condition known as Psoriasis, 4. A scaly condition of the scalp known as dandruff, 5. Acne, 6, Inflammation of the dermis and subcutaneous tissues known as Cellulitis usually due to an infection, 7. Boil or furuncle, 8. Skin abscess 9. Rosacea (A chronic skin condition causing a red rash), 10. viral infections causing excessive skin growth known as Warts, 11. Skin cancers, Melanoma, 12. Common skin cancer called Basal cell carcinoma, 13. Seborrheic keratosis (benign, often itchy growth that appears like a "stuck-on" wart), 14. Actinic keratosis (crusty or scaly bump that forms on sun-exposed skin), 15. Squamous cell carcinoma, 16. Herpes blisters on lips and genitalia caused by the herpes viruses HSV-1 and HSV-2, 17. Raised, red, itchy patches on the skin called hives arising due to skin allergies, 18. Tinea versicolor due to benign fungal skin infection, 19. Red rash affecting large areas of the skin especially in children called viral exanthema, 20. Painful rash on skin due to past infections by the chicken pox virus known as Shingles (Herpes zoster).

21. Scabies caused by tiny mites that burrow into the skin. 22. Ringworm due to a fungal skin infection, 23. Athlete's foot caused by certain fungal infections, 24. Vitiligo or Leukoderma (auto immune disease) involving white patches on skin due to loss of pigmentation.

Treatment options are 1. Skin surgery including skin grafts 2. Antihistamines 3. Corticosteroids, 4. Anti-fungal, anti-viral, anti-bacterial, anti-lice and other antimicrobials 5. Sunscreen oils and creams 6. Immune modulators, 7. Laser and heat treatments 8. Skin moisturizers 9. Skin bleaching agents, 10. Herbal treatments.

The following plant herbals are being recommended and considered useful for treating various skin ailments. These are based on anecdotal experience and in most cases have limited clinical data.

***Allium cepa* (Onion), *A. sativum* (Garlic): Kingdom Plantae, Division Angiospermae, Monocots, Order: Liliales, Family: Liliaceae)**

Extracts of onion and garlic might be promising in treatment of fungal-associated diseases from important pathogenic genera like Candida, Malassezia and the dermatophytes. Further, creams having Onion and garlic extract improved scar softness, redness, texture, and global appearance at the excision sites of surgery. In other studies, Garlic ingestion delayed formation of skin papilloma in animals and simultaneously decreased the size and number of papilloma, which was also reflected in the skin histology of the treated mice.

Aloe (*Aloe vera*): Kingdom: Plantae, Division Angiospermae, Eudicots, Order: Asperagales, Family: Asphodelaceae (formerly, Xanthorrhoeaceae)

Extracts of Aloe leaf are highly mucilaginous and are added to various moisturizing creams and lotions. It is also added to facial tissues as a soothing agent. The plant is rich in many polysaccharides including mucopolysaccharides. One of the ingredients **Acemannan** is an immune booster. Other ingredients include **Anthraquinones** (Alloin A and B) which manage the laxative properties of Aloe juice.

Ammi/Khella/toothpick weed (*Ammi visnaga*): Kingdom Plantae, Division Angiospermae, Eudicots, Order: Apiales, Family: Apiaceae: This herb growing mainly in the middle east with umbellate flowers has two important phytochemicals namely visnaga and Khellin. Khellin/Kheltin or extracts of Ammi is applied to the skin topically and exposed to light to stimulate production of melanocytes in vitiligo patients. In addition, extracts of Ammi breaks up the formation of kidney stones. It is also a smooth muscle relaxant and was used to control cardiac arrhythmia. But the plant extracts have severe side effects including potential for skin cancer and cardiac arrests. However, its use still prevails in the middle east.

Apaamaarga/Prickly chaff flower/Devil's horsewhip (*Achyranthes aspera*): Kingdom Plantae, Division Angiospermae, Eudicots, Order: Caryophyllales, Family: Amaranthaceae. Apaamaarga has been extensively used in Ayurveda as an anti-inflammatory agent besides being useful in treating hemorrhoids, indigestion, cough, asthma, anemia, jaundice, and snake bite. It is particularly recommended for control of dropsy (edema/swelling) wherein the leg, calf and or ankle are swollen due to fluid retention. The fluid retention may have many causes such as venous insufficiency, deep vein thrombosis and kidney/heart diseases. In some parts of India, the leaf juice is smeared on the skin of people stung by scorpions.

Ashoka tree (*Sarco asoca/Saraca indica*): Kingdom: Plantae, Division: Angiospermae, Eudicots, Order: Fabales, Family: Caesalpineaceae.

The Ashoka tree is considered a holy tree in India. Paste of the roots and leaves is prescribed for treating freckles and external inflammations, ulcers, and skin diseases. It is used for treating itching in eczema, psoriasis, dermatitis and herpes-kushta/visarpa by rubbing the crushed flower on the skin.

The plants are shrubs or short trees growing in tropical climes. They have glabrous serrated simple leaves. The red/yellow/pink or near white flowers are steeped in oil and the oil having flower extracts are used as a body oil, hair oil and as hair conditioners to treat dandruff and scalp disorders. The mucilaginous leaf extract is used as a skin salve or as a scalp treatment known as Thaali in India.

Babchi (*Psoralea corylifolia*): Kingdom: Plantae, Angiospermae, Eudicots, Fabales, Fabaceae: This plant is a native of Indian subcontinent. It is an annual plant about 3 ft tall. It produces pale violet flowers and after fertilization produce one seed per plant. The seeds are rich in coumarins including the DNA crosslinking agent Psoralen. In addition, the seeds have many other phytochemicals including flavanoids and meroterpenes. The seed and plant extracts are used to treat dermatitis, psoriasis, and vitiligo. The psoralen binds to thymine in DNA forming thymine dimers and thus preventing DNA replication especially when applied in combination with UV phototherapy. Babchi seed extracts are one of the main constituents of herbal medication formulations for treating **vitiligo**.

Balloon vine/Valli Uzhinja (*Cardiospermum halicacabum*): Kingdom: Plantae, Division Angiospermae, Eudicots, Order; Sapindales, Family: Sapindaceae. Powdered bark and extracts are internally administered for treating skin diseases, asthma, sore throat, diarrhea, and abdominal discomfort and applied externally for skin ulcers. In the skin, studies showed significant reduction in papilloma. The extracts from plant are added to creams and oils for treating eczema of the skin and dermatitis. A gel supplement known as Cardiospermum gel is being marketed as a treatment option for eczema. The herbal extracts are additives of hair oil preparations to reduce dandruff and for darkening the hair. The plant extract has anti-inflammatory and analgesic properties. It is used in treating earaches and in relieving swellings. Leaf blade is also combined with castor oil and applied to treat boils and swollen limbs. Like other herbs, it has many other attributes.

Burdock (*Arctium lappa*/A. sp.), Kingdom: Plantae, Division Angiospermae, Eudicots, Order: Asterales, Family: Asteraceae.

Burdock plants are biennial herbs with purple flowers. The flowers, bark, and leaf extracts are used to control skin conditions and are used in Chinese medicine as well as in American Indian folk medicine. Burdock root oil known as bur is used to control scalp infections.

Calendula/Pot marigold/Scotch marigold (*Calendula officianalis*): Angiospermae, Eudicots, Order: Asterales, Family: Asteraceae.

Calendula is a perennial herb belonging to the Sunflower family. Leaves and flower extracts are prescribed for everything from acne to wounds and helps to reduce inflammation and promote cell repair. It can also be brewed as a tea and used as a compress, wash, or poultice. Flowers have carotene Zeaxanthin and lutein that are considered good for eye health. Besides, they have, saponins, flavaxanthin, other carotenoids and essential oils that are components in cosmetic skin preparations.

Cannabis/Marijuana/Charas/Ganja/pot (*Cannabis sativa*): Kingdom Plantae, Division Angiospermae, Eudicots, Order: Rosales, Family: Cannabaceae.

Hemp oil having cannabidiol (CBD) applications control pain associated with certain skin conditions including psoriasis. Hemp seed oil with CBD is useful for treatment of eczema and host of other skin diseases like dermatitis, seborrheic dermatitis/cradle cap, varicose veins, eczema, psoriasis, lichen planus and acne rosacea.

Chickweed (*Stellaria media*): Kingdom: Plantae, Division Angiospermae, Eudicots, Order: Caryophyllales, Family: Caryophyllaceae.

Chickweed plants are annual herbs found in lawns, meadows, and other forage areas. Plants have ovate glabrous leaves with small flowers with sepals, petals stamens and carpels in groups of 5. Infusions and pastes of the stem and leaves are used to treat skin infections and abrasions.

Comfrey (*Symphytum uplandica or S. officinale*): Kingdom: Plantae, Division Angiospermae, Eudicots, Order: Boraginales, Family: Boraginaceae

Comfrey is an herb with simple alternate leaves. Comfrey leaf extracts are used to treat skin injuries. It has pronounced wound healing properties and hence earned the nickname "knitbone" because it is said to 'knit' wounded tissues back together. The plant has Allantoin (wound healing chemical), saponins, pyrrolodizine alkaloids, mucilage, and other phytochemicals. Although the leaf extracts were prescribed for internal consumption for wound healing, it turns out that it is extremely toxic. Hence, the USA FDA has issued warnings about internal use and warning labels must be inserted even for topical use. Pregnant women should not use Comfrey.

Curly Dock/Yellow Dock Root (*Rumex crispus*): Kingdom: Plantae, Division Angiospermae, Eudicots, Order: Caryophyllales, Family: Polygonaceae,

Powdered roots are applied as a poultice to control itching, Athletes foot fungus and skin sores.

Echinacea angustifolia, E. purpurea (Purple coneflower): Kingdom: Plantae, Division Angiospermae, Eudicots, Order: Asterales, Family: Asteraceae, Family: Asteraceae.

Echinacea has been applied to and used to treat skin problems such as skin boils, wounds, ulcers, burns, herpes, hemorrhoids, and psoriasis.

Helichrysum/Italian strawflower, (*Helichrysum italicum, Orientale*), Kingdom: Plantae, Division Angiospermae, Eudicots, Order: Asterales, Family: Asteraceae.

This is an herb growing in Mediterranean countries, Madagascar, Australia, and certain other African and Asian countries. Over 60 species belong to this genus. *Helichrysum italiae* is the most known for its skin-healing and anti-aging essential oil. The essential oil is obtained by steam distillation of the leaves and flowers. The oil by itself or as an ingredient of facial and skin creams is marketed for its presumed anti-ageing properties. The oil is applied to the face daily to target fine lines and wrinkles. Helichrysum oil is used to help promote healthy skin cell regeneration and a glowing, youthful complexion. Topical application of water extracts of the plant relieves skin itch, boils, carbuncles, and skin inflammations.

Henna: (*Lawsonia inermis*): Kingdom: Plantae, Division Angiospermae, Eudicots, Order: Myrtales, Family: Lythraceae.

Leaf extracts applied as a paste known, as Henna is a traditionally used cosmetic preparation, which gives a red/orange color to the skin and seems to have anti-bacterial and antifungal properties. Henna is also used as a hair color agent.

Hibiscus/Shoe flower (*Hibiscus rosasinensis*), Kingdom: Plantae, Division Angiospermae, Eudicots, Order: Malvales, Family: Malvaceae.

Extract of the flower are considered to have UV protecting properties and are added to skin creams for this purpose. Flower extracts are also added to cold drinks as flavoring agents. Leaf extracts which are mucilaginous are applied to skin and hair in preparations called "Thaali" to improve the luster of skin as well as contribute to hair growth. Hibiscus leaf and flower extracts in oil called Chembaruthi oil/shampoo/conditioners/body lotions are Ayurvedic preparations prescribed for toning up skin, prevent skin diseases and improve hair growth.

Hot pepper (*Capsicum annum*): Kingdom: Plantae, Division Angiospermae, Eudicots, Order: Solanales, Family: Solanaceae). Family: Solanaceae. 0.025% Capsaicin from capsicum plants are components of creams used to control pain associated with psoriasis. It is reported to not only control pain but also reduce itching.

Lavender (*Lavendula angustifolia*) Order: Kingdom: Plantae, Division Angiospermae, Eudicots, Order: Lamiales, Family: Lamiaceae. The herb, essential oil, hydrosol, and infused carrier oils are all useful for skin preparations. Lavender buds can be infused into carrier oils, creams, and other skin aids.

Mama Cadela/Sweet Cotton (*Brosimum gaudichaudii*): Kingdom: Plantae, Division Angiospermae, Eudicots, Order: Rosales, Family: Moraceae. The plants are small trees related to bread fruit and figs. The plants are natives of Brazil and nearby south American countries. It is usually sold in the form of its sliced roots. The active ingredient in this plant that helps to treat de-pigmentation patches is called bergapten which are furocoumarins. Furocoumarins are potent, organic compounds produced by plants as a part of their defense mechanism. The bark and fruits have this compound giving it a bad taste. Extracts of this plant delay or stop the destruction of the melanocyte pigment cells so that people who are affected by vitiligo get relief. The leaves or roots are boiled along with some water. A compress made of soft cloth is dipped in the cooled decoction and applied on the vitiligo patches. This can be done up to twice a day for at least two weeks to gauge if the therapy is having any appreciable effect.

Neem (*Azadirachta indica*): Kingdom: Plantae, Division Angiospermae, Eudicots, Order: Sapindales, Family: Meliaceae.

Neem leaf paste is traditionally used to counter all sorts of skin conditions. The oil is used topically while the leaf or powdered herb is taken orally. The plant has two anti-inflammatory and antibacterial compounds called **nimbidol and gedunin.** Neem oil and leaf extracts are effective against fungi, bacteria and insects and are used to treat Eczema and psoriasis. The oil also protects from lice in hair. In Ayurveda and Siddha medicine, Neem products are used for healing ulcers and skin imperfections caused by the aftereffects of small and chicken pox.

Neem trees are fast growing plants native to the Indian sub-continent and now grown in most tropical and sub-tropical climes. Plants bear serrated compound leaves and small white flowers, which may be protandrous (male to female), bisexual or male. Fruits are drupes with a thin green or pale yellow exocarp, pulpy mesocarp and an endo carp having seeds. The leaves have oil glands and Neem oil is obtained by crushing the fruits with seeds. Neem oil and leaf extracts (Bio-insecticides) are now commonly used in organic farming to control insects affecting crop and horticultural plants.

Pokeweed/Poke root (*Phytolacca americana*), Kingdom: Plantae, Division Angiospermae, Eudicots, Order: Caryophyllales, Family: Phytoloccaseae.

Pokeweed is a perennial plant with simple alternate leaves. Both fruit berries and leaf extracts are used in American Indian medicine as a treatment for skin problems to treat scabies, psoriasis, ulcers, and varicose veins. The FDA does not endorse these claims although there are various supplements that are marketed.

Rosemary (*Rosmarinus officinalis*): Kingdom: Plantae, Division Angiospermae, Eudicots, Order: Lamiales, Family: Lamiaceae/Labiatae. Rosemary is a common household plant grown in many parts of the world. It is used for flavoring food, a beverage drink, as well as in cosmetics. Aqueous extract of R. officinalis has been reported to be effective in preventing cutaneous photo damage induced by UV radiations. Rosmarinic acid also controls acne.

Saffron (*Crocus sativus*): Kingdom: Plantae, Division Angiospermae, Monocots, Order: Aspergales, Family: Iridaceae.

Saffron is a naturally derived plant product from the stamens and dried petals that has multiple medicinal properties. The chemo preventive effect of aqueous saffron on chemically induced skin carcinogenesis using a histopathological approach was studied. Its ingestion inhibited the formation of skin papilloma in animals and simultaneously reduced their size. In popular culture in India, it is believed that intake of saffron by pregnant women will result in Children being born with lighter skin color. There is no scientific basis for this. Crushed whole plant extracts applied externally with a pinch of salt controls ringworm. Ringworm is not an animal, but a fungal disease of the skin closely related to jock itch and athlete's foot fungus. The stamens are dried and used to flavor foods. It is also used in skin-health preparations.

Turmeric/Manjal/Kumkum (*Curcuma longa*}: Kingdom: Plantae, Division Angiospermae, Monocot, Order: Zingiberales, Family: Zingiberaceae. Plant details are given elsewhere in chapter on Digestive conditions. Turmeric is widely used in Ayurvedic medicine to keep inflammation in check. The rhizomic roots are ground into a paste and applied on skin to control skin infections, sores, cracked feet and for general cosmetic. It is a part of some Ayurvedic skin creams. The active part of turmeric curcumin is an antioxidant. Plant extracts are used to treat boils, scabies, and eruptions of skin and other skin diseases.

 Witch Hazel (*Hamamelis* sp.), Six species namely, *H. Mexicana, H. Ovalis, H. Virginiana, H. vernalis, H. Japonica* and *H. Mollis* are collectively referred as of witch hazels. Order: Saxifrages, Family: Hamameliceae (Saxifragaceae)

Witch-hazel extracts are folk remedies for treating psoriasis and eczema, ingrown nail applications, insect bites and poison ivy. Extracts are ingredients of aftershave lotions.

Other Herbs

Extracts of Dandelion flowers and leaves, Elderberry flowers, Borage, Chamomile, Parsley, Rosemary, Thyme, Yarrow (*Achillea millefolium*), Marshmallow (*Althaea officinalis*), carrot seed oil, Rose petal oil, Plantain leaf extracts and St John's wort aerial parts are all components of various skin and hair supplements. Mucilaginous extracts from English plantain (*Plantago lanceolate*), fenugreek (*Trigonella foenum-grecum*), mullein (*Verbascum thapsus*), slippery elm (*Ulmus fulva*), and flax (*Linum usitassimum*) contain mucilage's, which act as emollients and soothe the skin.

References

https://www.ncbi.nlm.nih.gov/books/NBK92761/

https://www.ncbi.nlm.nih.gov/pmc/articles/PMC3931201/

https://www.researchgate.net/publication/260983682_pharmacological_and_medicinal_uses_of_achyranthes_aspera

Agrawal RC, Pandey S. (2009): Evaluation of anticarcinogenic and antimutagenic potential of Bauhinia variegata extract in Swiss albino mice. Asian Pac J Cancer Prev. 2009; 10: 913–6.

Arora N, Bansal MP, Koul A. (2011): Azadirachta indica exerts chemopreventive action against murine skin cancer: Studies on histopathological, ultra-structural changes and modulation of NF-kappaB, AP-1, and STAT1. Oncol Res. 2011; 19: 179–91.

Bedi M.K, Shenefelt P.D., (2002): Herbal therapy in dermatology. Arch Dermatol. 2002; 138: 232–42.

Beylier M.F. (1979): Bacteriostatic activity of some Australian essential oils. Perfum Flavourist. 1979; 4(2): 23–5.

Chakraborty A, Brantner A, Mukainaka T, Nobukuni Y, Kuchide M, Konoshima T, et al. (2002): Cancer chemopreventive activity of Achyranthes aspera leaves on Epstein-Barr virus activation and two-stage mouse skin carcinogenesis. Cancer Lett. 2002, 177: 1–5.

Chan B.C, Hon K.L, Leung P.C, et al., (2008): editors. Traditional Chinese medicine for atopic eczema: Pent Herbs formula suppresses inflammatory mediators release from mast cells. J Ethnopharmacol. 2008, 120(1): 85–91.

Fonseca YM, Catini CD, Vicentini FT, Nomizo A, Gerlach RF, Fonseca MJ, (2010): Protective effect of Calendula officinalis extract against UVB-induced oxidative stress in skin: Evaluation of reduced glutathione levels and matrix metalloproteinase secretion. J Ethnopharmacol. 2010; 127: 596–601.

Ghazanfar S.A. (1994): Handbook of Arabian Medicinal Plants. Boca Raton, FL: CRC Press.

Jowkar, Jamshidzadeh, A., Mirzadeh Yazdi, A, and Pasalar, M (2011), The Effects of Fumaria Parviflora L Extract on Chronic Hand Eczema: A Randomized Double-Blind Placebo Controlled Clinical Trial, Iran Red Crescent Med J. 2011 Nov; 13(11): 824–828, PMCID: PMC3371888, PMID: 22737422.

Kapoor L.D. (1990): CRC Handbook of Ayurvedic Medicinal Plants. Boca Raton, FL: CRC Press.

Kingston C, Jeeva S, Jeeva GM, Kiruba S, Mishra BP, Kannan D. Indigenous knowledge of using medicinal plants in treating skin diseases in Kanyakumari district, Southern India. Indian J Tradit Knowl. 2009; 8: 196–200.

Renu, S. (2010): Treatment of skin diseases through medicinal plants in different regions of the world. Int J Compr Pharm. 2010; 4: 1–4.

Tabassum, N., and Hamdani, M, M. (2014): Plants used to treat skin diseases. hPharmacogn Rev. 2014 Jan-Jun; 8(15): 52–60.doi: 10.4103/0973-7847.125531, PMCID: PMC3931201, PMID: 24600196.

Vaughn, A. R.; Branum, A; Sivamani, RK (2016). "Effects of Turmeric (Curcuma longa) on Skin Health: A Systematic Review of the Clinical Evidence". Phytotherapy Research. 30 (8): 1243–64. doi: 10.1002/ptr.5640. PMID 27213821.

CHAPTER 10

Herbals for ENT and Respiratory Systems

Eye

Herbs have been used for thousands of years to cure illnesses of the eye. Good Eye health is affected by cataracts, red eye, dry eye, conjunctivitis, glaucoma, retinitis, retinopathy, macular degeneration, certain cancers and infections The following is a list of herbs that are used by practioners of traditional medicine for maintaining good eye health. Good clinical studies for using these herbals are meager.

Bilberries/European Blue berries/Whortleberry (*Vaccinium sp.*): Kingdom: Plantae, Clade: Angiospermae, Eudicots, Order: Ericales, Family: Ericaceae As per folklore and traditional medicine, extracts of the fruits are said to improve night vision but there is no medical evidence to back these claims. Extracts of this herb is used, as compress, tea, or poultice. It is a traditional herbal medicine used in Europe to sooth itchy eyes and conjunctivitis.

Eyebright (*Euphrasia rostkoviana/E. officinalis.*): Kingdom: Plantae, Clade: Angiospermae, Eudicots, Order: Lamiales, Family: Orobanchaceae.

Whole plant extracts are prescribed as a tea or as a compress for treating Eye diseases.

Fennel (*Foeniculum vulgare*): Kingdom: Plantae, Clade: Angiospermae, Eudicots, Order: Apiales, Family: Apiaceae

Fennel seed extract is said to be particularly helpful for treating watery and inflamed eyes. Apart from these mild conditions, Fennel can also be used to treat cataracts and glaucoma.

Grape seed (*Vitis vinifera*): Kingdom: Plantae, Clade: Angiospermae, Division: Eudicots, Order: Vitales, Family: Vitaceae.

The extract of this seed is rich in phytochemicals that are good for the eyes. It also has antihistamine and antioxidant properties.

Saffron (*Crocus sativus*): Kingdom: Plantae, Clade: Angiospermae, Division: Eudicots, Order: Asperagales, Family: Iridaceae

Extracts of Saffron flower l is used to treat cataracts. Furthermore, it can also delay degeneration of eyesight among elderly people. This herb can reduce the risks of glaucoma and macular degeneration by acting as a cerebro-spinal dilator.

Turmeric (Curcuma *longa*): Kingdom: Plantae, Clade: Angiospermae, Division: Eudicots, Order: Zingiberales, Family: Zingiberaceae.

Turmeric shows many properties including control of inflammation. Intake of turmeric is said to reduce the oxidation of the eye lens.

The Ear

The ear is the organ of hearing and balance. The ear has three parts—the outer ear, the middle ear, and the inner ear. The outer ear consists of the pinna and the ear canal. The middle ear includes the tympanic cavity and the three ossicles. The inner ear sits in the bony labyrinth, and contains the semicircular canals, which enable balance and eye tracking when moving; the utricle and saccule, which enable balance when stationary; and the cochlea, which enables hearing. The ear may be affected by disease, including infection and traumatic damage. Diseases of the ear may lead to hearing loss, ear pressure, ringing, and balance disorders such as vertigo. Breathing disorders, sinus pressure, and infections of not only ear but of nose, sinuses, and throat will affect ear health. Apart from hygiene and good nutrition, many herbals are also considered to be of use in keeping good ear health.

Herbs for Ear Health

Folic acid is extremely important to ear health. According to a controlled study conducted in the Netherlands, after three years, those who got folic acid pills had less low-frequency hearing loss than did placebo recipients. Folic acid is present in most green vegetables as well as in many herbs. Likewise, Zinc boosts the body's immune system and manages cell growth and healing wounds, Zinc can protect against ear infections, and promotes healthy cell growth in ears and throughout body. Various studies showed that Zinc supplementation helped control of tinnitus, balance, and hearing loss among elderly patients. Some studies suggest that zinc is also effective in treating tinnitus in individuals with normal hearing.

Foods with Zinc include Dill, Spinach, cashews, almonds, peanuts, beans, split peas, lentils, and dark chocolate. Zinc is an important nutrient for hearing health.

Garlic (*Allium sativum*): Kingdom: Plantae, Clade: angiospermae, Clade: Monocots, Order: Alliales, Family: Amaryllidaceae.

Garlic oil has anti-bacterial properties and thus can control ear infections. The oil can be applied as drops into the ear.

Gingko (*Ginkgo Biloba*): Kingdom: Plantae, Clade: Tracheophytes, Order: Ginkgoales, Family: Ginkgoaceae.

Gingko can promote better blood circulation. Proper blood circulation to your ears keeps everything functioning and keeps ears at a safe temperature.

Turmeric (*Curcuma longa*): Kingdom: Plantae, Clade: angiospermae, Clade: Monocots, Order: Zingiberales, Family: Zingiberaceae.

Turmeric is related to ginger and has been used for centuries as a cure for viral infections, inflammatory conditions, dermal health, and many other health related conditions. Turmeric is also considered to help control ear conditions and inflammations in the ear.

Others

Ginger tea, Tea tree oil and Cajeput essential oil are also being recommended by the supplements industry for Ear health. However, there are no scientific publications to support these claims.

Allergic Rhinitis

Allergic rhinitis is an inflammatory condition that affects the nasal passages, sinuses, ears, and throat. Also known as hay fever, it occurs when a person inhales an allergen to which they are sensitive. There are two types of allergic rhinitis namely one that is caused seasonally known as seasonal allergy and the other occurring perennially caused most often by house dust, pet hair, mites, cooking smells, oil smoke, and molds. A third kind is genetic and is triggered by any factor including stress.

Herbals for Allergies

Stinging Nettle (*Urtica dioica*): Kingdom: Plantae, Clade: angiospermae, Clade: Eudicots, Order: Urticales, Family: Urticaceae

Nettles are herbaceous perennial plants growing wild in temperate climes. The leaves and stem are covered with hollow stinging hairs that act like injection syringe needles that inject histamine upon contact resulting in contact dermatitis. Cooking the leaves will remove their stinging effects, and cooked leaves can be added to salads, soups, or stews, like most other green leafy vegetables. Nettle leaves are used by traditional medicine practitioners for treating kidney, gastrointestinal, rheumatism, and allergic disorders.

Perilla (*Perilla frutescens*): Kingdom: Plantae, Clade: angiospermae, Clade: Eudicots, Order: Lamiales, Family Lamiaceae.

This annual herb grows well in the Korean peninsula, and South East Asia including India. It is a member of the mint family. The variety *Perilla frutescens var crispa* is grown in Japan and used to control fish and crab meat allergies. It can help in the battle against allergic rhinitis symptoms also. The plant is toxic to cattle. The essential oils found in Perilla have an antidepressant effect and boost serotonin levels in the brain. Many studies have shown Perilla to be useful for the treatment of nasal congestion, sinusitis, allergic asthma, and eye irritation (another trouble for many allergy sufferers). The leaves have **perilla ketone, egoma ketone, and isoegoma ketone, perillaldehyde, limonene, linalool, beta-caryophyllene, menthol, and alpha-pinene.** The leaves are toxic to cattle and cause acute respiratory syndrome or panting disease. So, although the seed extracts, leaves are used in culinary preparations as also to control allergies, extreme caution is advised in view of the known toxicities in cattle.

Sea Buckthorn (*Hippophae rhamnoides*): Kingdom: Plantae, Clade: angiospermae, Clade: Eudicots, Order: Rosales, Family: Elaeagnaceae.

Sea buckthorn has many nutrients and phytochemicals. This extremely nutrient-dense berry offers an array of organic acids, tannins, **quercetin, Provitamin A**, vitamin E, and a great deal of vitamin C, as well as B complex vitamins. Sea buckthorn is prescribed for people who suffer from allergic rhinitis, as well as asthma, chronic coughs, and other breathing disorders. There are many sea buckthorn products on the market today.

Butterbur (*Petasites hybridus*): Kingdom: Plantae, Clade: angiospermae, Clade: Eudicots, Order: Asterales, Family: Asteraceae.

The butterbur shrub grows in the marshes of North America, Europe, and Asia. People have used it traditionally for the treatment of pain, headaches, fevers, and digestive ailments. More recently, it has also been used for urinary tract infections, headaches, including migraines, as well as for the treatment of hay fever. The parts used are rhizomes, root, or leaves, which are extracted and prescribed as herbal medicine. **Toxicity:** It is toxic to liver and may cause cancers. The product used for allergic rhinitis should be from a well-known source and should not be from wild growth.

Yarrow (*Achillea millefolium*): Kingdom: Plantae, Clade: angiospermae, Clade: Eudicots, Order: Asterales, Family: Asteraceae.

Yarrow is a perennial herb native to the British Isles, but also common throughout Europe and Asia. This multi-purpose herb has antiseptic, stomachic, antispasmodic, astringent, and diaphoretic properties. Used traditionally to treat colds, flu, and fevers, yarrow can also be a useful remedy against allergic rhinitis. The essential oil from this plant is blue black in color and is toxic to mosquito larvae.

Herbs for Sinus Relief

Extracts of leaf of Mullein (*Verbascum thapsus*), Chamomile tea, Comfrey, Marshmallow, Ginger, Eyebright, Fenugreek (*Foeniculum foenum-grecum*), Thyme (*Thymus* sp), Echinacea (*Echinacea* sp.), Eucalyptus oil, Menthol, Oil of Wintergreen, *Calendula* flowers and leaves (*Calendula officianalis*), Astragalus and Horseradish are all being recommended as herbals for sinus relief. However, there is no scientific verification to suggest that any of these are of value in controlling sinus issues.

Nose Throat and Respiratory Systems

The human and other mammalian respiratory system consists of: 1. the upper respiratory tract made up of the nose with the two nostrils, the sinuses, pharynx, and the upper part of the larynx (voice box). 2. The lower respiratory tract consisting of the lower part of larynx, trachea, the two bronchi which branch into the lungs and get divided into bronchioles and alveoli where exchange of gases take place. The lungs are protected and covered by the pleural membrane forming the pleural cavity lined with lubricating fluid so that the expansion and contraction of the lungs occurs smoothly. Disruptions in the normal functioning of the respiratory pathways can occur due to: blockages, nasal and sinus inflammations, inflammation of the pharynx (pharyngitis), larynx (laryngitis), bronchitis, pleurisy, asthma due to allergies, primary and secondary lung cancers, mesothelioma (asbestos), coal inhalation, silicosis, and chronic obstructive pulmonary disease. The cause for any of these disruptions could be allergies, bacterial, viral, mycoplasma and fungal infections, particulate matters such as breathing coal dust resulting in coal miners lungs, breathing silica particles causing silicosis and asbestos particles/fibers causing mesothelioma. Other causes could be smoking and other environmental causes. Finally, yet importantly are genetic causes resulting in cancers.

Certain herb extracts and formulations have been in use by traditional herbologists for improving and treating some of these respiratory conditions.

Adathoda/Malabar nut/Vasaka (*Justicia vasica syn Adhatoda vasika*): Kingdom: Plantae, Angiospermae, Eudicots, Order: Lamiales, Family: Acanthaceae.

This shrub plant is found in the Indian subcontinent, South Asia, China, Malay Archipelago including Indonesia and in South China as well as in the Caribbean and equatorial South America. The plant extracts are widely prescribed in TCM, Ayurveda, Siddha and Unani medicine. Fresh and dry leaf extracts are used as expectorant, antispasmodic, and bronchodilator in tonics, expectorants, and syrups. It is a common part of many types of anti-asthma syrups. In the Ayurvedic system, the plant extracts are used in formulations for treating some types of fever, secondary effects of tuberculosis, dyspnea, and arthritis.

Angelica/Norwegian Angelica (*Angelica archangelica*): Kingdom: Plantae, Clade: Angiospermae, Eudicots, Order: Apiales, Family: Apiaceae.

Plants are biennials grown in Scandinavian and East European countries for extracting essential oil for flavoring and medicinal purposes. Plants have compound serrated leaflets with hollow stem. Plants grow vegetative for first year and produce small white-green flowers in umbels in second year of growth. Seeds are mericarps and are ridged like seeds of Dill. Seeds, roots, and stems are extracted to produce aromatic oil used to flavor alcoholic drinks like Gin, Absinthe and as medicine for treating respiratory and gastro-intestinal illnesses.

Beach Salvia (*Salvia africana-lutea*) Kingdom: Plantae, Angiospermae, Division: Eudicots, Order: Lamiales, Family: Lamiaceae.

These shrubby short plants grown in South Africa are used to treat coughs, colds, and sinus blockages by the native population.

Cocklebur (*Xanthium strumarium*): Kingdom: Plantae, Angiospermae, and Division: Eudicots, Order: Asterales, Family Asteraceae.

Cockleburs are annual herbs producing fruits known as "burrs". The fruit extracts are used in TCM, to treat sinus congestion, chronic nasal obstructions and discharges, and respiratory allergies. Xanthium is toxic and causes vomiting, diarrhea, and abdominal pain.

Echinacea/Purple cone flowers (*Echinacea angustifolia/E. purpurea*): Kingdom: Plantae, Clade: Angiospermae, Clade: Eudicots, Order: Asterales, Family: Asteraceae

Echinacea root and whole plant extracts are sold in many forms as immune boosters and in formulations to treat colds, cough, and other respiratory illnesses. It is a native American remedy considered a cure all. Plants are perennial herbs producing sunflower-like flowers. The extracts are hepatotoxic, and it should not be administered to patients with autoimmune diseases.

Elderberry (*Sambucus* sp.), Kingdom: Plantae, Class: Angiospermae, Eudicots, Order: Dipsacales, Family: Adoxaceae (Caprifoliaceae).

Sambucus are shrubby plants bearing white or off-white flowers in inflorescences and black or blue-black berries. Extracts of the berries are rich in flavanoids. Both flower and fruit extracts are cooked and processed before being used in cough syrups and expectorants for treating the common cold and flu-like symptoms. **Toxicity:** the berries have cyanogenic glycosides and so eating fresh un- cooked fruits or flower extracts cause nausea, vomiting and intestinal distress.

Ephedra/Mormon tea/Brigham tea (*Ephedra sp.*), *Kingdom*: Plantae, Class: Gymnospermae, Order: Ephedrales, Ephedraceae.

There are over 50 species of Ephedra growing in different parts of the world. Thus, *E. sinica* grows in China, Siberia, Mongolia, *E. nevadensis* and *E. viridis* in Arizona, California, Nevada, Utah, (USA), *Ephedra sumlingensis* in the western Himalayas, *E. dahurica* in Mongolia and Siberia etc. All are naked-seed gymnosperms with jointed green slender stems and scaly leaves arising from the nodes. They produce male and female cones instead of typical flowers. The plants have the alkaloids ephedrine and pseudoephedrine as well as several other phytochemicals. Ephedrine is a cardio stimulant and decongestant. It was a constituent of many cough expectorants, decongestants, and Asthma tablets, but is no longer used in allopathic formulations because of severe adverse effects on the heart. However, it is still available as a TCM drug as well as in herbal preparations in many countries. Ephedrine was also popular as a weight reduction and performance-enhancing drug, but its extensive use has serious consequences on cardio-health. In India, it is thought to be one of the mythical soma plants used in religious ritualistic drink ceremonies. It has hallucinogenic and broncho dilator properties. **Toxicity:** causes Tachycardia, vaso constriction, nervousness, hallucinogenic, flushing, difficulty in urination and addiction.

Eucalyptus/Tasmanian blue gum (*Eucalyptus globulus*): Kingdom: Plantae, Angiospermae, Eudicots, Order: Myrtales, Family: Myrtaceae.

A widely cultivated native tree of Australia, this tall plant has aromatic leaves and fruits that have medicinal value. The leaves are steam distilled to generate eucalyptus oil that is used as a chest rub to alleviate chest congestion and as a painkiller when rubbed on joints.

Garlic (*Allium sativum*) Kingdom: Plantae, Angiospermae, Monocots, Order: Asperagales, Family: Amaryllidaceae.

The garlic cloves and oil are considered good for treating the common cold but the evidence for this claim is minimal.

Ginger Lily (*Hedychium spicatum*; Syn *H. album*): Kingdom: Plantae, Angiospermae, Clade: Monocots, Order: Zingiberales, Family: Zingiberaceae.

Plants are herbaceous annuals with long green leaves with parallel veins. The plants grow in tropical and semi-tropical climes. Extracts of the rhizome of this relative of ginger is prescribed for treatment of Asthma and as an expectorant by Ayurvedic, Siddha, Unani and TCM practitioners. Like ginger, the rhizome extracts are also prescribed for other health conditions such as stomach pains, gastro-intestinal disorders, pain and even snake bites.

Hyssop (*Hyssopus officinalis*): Kingdom: Plantae, Angiospermae, Eudicots, Order: Lamiales, Family: Lamiaceae.

Hyssop is an herbaceous shrub with a woody branching stem bearing lanceolate leaves and achene fruits. The plant is mentioned in the Bible and has been in use as a cough and expectorant medicine since ancient times by the Greeks, Egyptians, Israelis, and people living in the middle east. The whole plant is chopped and dried before use in medicines and in bitter liquors as digestive stimulants. **Toxicity:** the essential oil can cause convulsions and epileptic attacks.

Kandankathiri/Thai striped nightshade (*Solanum Xanthocarpum/S. virginianum*): Kingdom: Plantae, Clade: Angiospermae, Clade: Eudicots, Order: Solanales, Family: Solanaceae.

This is a shrubby herb growing in South Asia. The roots of the plant are extracted, and a water decoction is prescribed for treatment of coughs and bronchial conditions. A recent study concluded that the fruits of this plant have ant-cancer activities.

Kani pepper/Senegal pepper/Ethiopian pepper, Moor pepper and African American pepper (*Xylopia aethiopica*): Kingdom: Plantae, Clade: Angiospermae, Clade: Magnolids Order: Magnoliales, Family: Annonaceae.

This evergreen tree is a native of Africa Ethiopia, Senegal, Ghana, Nigeria, Ivory coast, and other countries in Africa. The bark of the tree steeped in water or wine is prescribed as a treatment for Asthma and bronchial infections. The dried fruits are soaked in water to produce a tea-like medicinal drink and fruit pods are used to spice up soups and other culinary items.

Mexican mint/Indian Borage/Panikoorkka/Karpooravalli (*Plectranthus amboinicuse*) Kingdom: Plantae, Clade: Angiospermae, Clade: Eudicots, Order: Lamiales, Family: Lamiaceae.

Plants are perennial herbs with fleshy aromatic leaves. The leaves and leaf extracts are used to treat cough and sinus congestions and as an expectorant by traditional herbologists in the Indian sub-continent. The leaves that have an oregano-like flavor are used as a culinary herb.

Mullein/Aarons rod (*Verbascum Thapsus*): Kingdom: Plantae, Clade: angiospermae, Clade; Eudicots, Order: Lamiales, Family: Scrophulariaceae.

Plants are bi-annual herbs with tall stems arising from a rosette of leaves and bearing an inflorescence with small yellow flowers. Mullein extracts of leaves and flowers have emollient properties and are components of cough syrups and mixtures to treat catarrh and pulmonary congestions. The extracts are added to skin treatment creams and root extracts have anti-athlete's foot fungus properties.

Opium poppy (*Papaver somniferum*): Kingdom: Plantae, Clade: angiospermae, Clade: Eudicots, Order: Ranunculales, Family: Papaveraceae.

Details of this plant are given in the Chapter on pain and analgesics. The resin from the poppy flower buds is the source for Codeine and Morphine. Codeine is a part of many cough suppressant medications. However, current investigations do not support that codeine has any substantive cough-suppressant effects. On the contrary, because of side effects such as constipation, feeling of inertia and general somnolence, it is not recommended in cough medicines for children and only doubtfully so for adults.

Thyme (*Thymus vulgaris*): Kingdom: Plantae, Clade Angiospermae, Clade: Eudicots, Order: Lamiales, Family: Lamiaceae.

This common culinary herb has the essential oil **Thymol** that has antibacterial properties. Thymol is a constituent of many mouthwashes and is recommended as a gargling agent for treating inflamed throat muscles.

Turkey Bush (*Eremophila gilesii*): Kingdom: Plantae, Clade: Angiospermae, Clade: Eudicots, Order: Lamiales, Family: Scrophulariaceae,

Plants are endemic to Australia and are bushy prostrate shrubs. The plant has a resin that has medicinal value in native Australian bush medicine. Resin and plant extracts are used for treating respiratory illnesses such as colds and associated headaches and pains.

Turmeric and white turmeric (*Curcuma longa and C. zedoaria*): Order: Zingiberales, Family: Zingiberaceae.

Both yellow and white turmeric rhizomes and their powders are steeped in water and used as throat gargles for treating sore throats and as hot-water drinks for treating the common cold. The evidence is mostly anecdotal.

Other Herbs

The following herbs are also sources for formulations for treating different respiratory conditions. They are:

Long pepper (*Piper longum*), Bibhitaki (*Terminalia bellirica*), Chebulic myrobalan (*Terminalia chebula*), Indian Sarsaparilla (*Hemidesmus indicus, Syn Periploca indica Linn*), Kizha Nelli (*Phyllanthus niruri*), Black pepper (*Piper nigrum*), Ginger (*Zingiber officinalis*), Holy Basil (*Ocimum sanctum*), Agnidamani (*Solanum Trilobatum*), Lily bulbs (*Bai He, Lilium sp*), Bitter Kola (*Garcinia kola*), Licorice plant (*Helichrysum petiolare*), African Teak (*Milicia excelsa*), Orange climber (*Toddalia asiatica*), Paper bark acacia (*Vachellia sieberiana*), Ipecac (*Cephaelis ipecacuanha*), Putty root (*Aplectrum hymale*); Indian tobacco (*Lobelia inflate*), Maiden hair fern (*Adiantum Capillus-veneris*); Dogwood, (*Cornus florida, syn. Benthamidia florida*), Willow (*Salix sp*), Caraway/Kala Jeera/ Perum Jeerakam (*Carum carvi*), Musk Okra (*Hibiscus abel-moschus, Syn*

A.moschatus), Ajwain/Carum (*Trachyspermum ammi, T. copticum*) and Anise (*Pimpinella anisum*).

References

https://en.wikibooks.org/wiki/Home_Remedies/Echinacea

https://en.wikibooks.org/wiki/Traditional_Chinese_Medicine/Usage_Of_Single_Herbs

https://www.audicus.com/herbs-hearing-loss/

https://www.healthline.com/health/reverse-hearing-loss#home-remedies

https://www.ncbi.nlm.nih.gov/pubmed/3492920

Bunney, S. (1984): Sea buckthorn". The Illustrated Encyclopedia of Herbs: Their Medicinal and Culinary Uses. Edited by Sarah Bunney, Dorset, 1984, p. 166.

Kowalchik, C and Hylton, W.H., (1987): Ginger." Rodale's Illustrated Encyclopedia of Herbs. Edited by Claire Kowalchik and William H. Hylton, Rodale Press, 1987, p. 223.

Lipkowitz, Myron A., and Tova Narava. The Encyclopedia of Allergies. Facts on File, Inc., 2001 (second edition), pp. 13-14.

Person OC, Puga MES, da Silva EMK, Torloni MR. (2016): Zinc supplementation for tinnitus. Cochrane Database of Systematic Reviews 2016, Issue 11. Art. No.: CD009832. DOI: 10.1002/14651858.CD009832.pub2

Schapowol A. and Petasites Study Group, "Randomised Control Trial of Butterbur and Cetirizine for Treating Seasonal Allergic Rhinitis," BMJ Vol. 324, no. 7330, 2002, pp. 144-46.

Schoffro Cook, Michelle. Allergy-Proof Your Life: Natural Remedies for Allergies That Work! Humanix Books, 2017.

Shambaugh George E. Jr. M.D. (1986): Zinc for tinnitus, Imbalance, and hearing loss in the elderly: The American Journal of Otology: November 1986 - Volume 7 - Issue 6 - ppg 476-477.

Shashank Kumar, and Pandey, A. K (2014): Medicinal attributes of Solanum Xanthocarpum fruit consumed by several tribal communities as food: an in vitro antioxidant, anticancer and anti-HIV perspective, BMC Complement Altern Med. 2014, 14: 112. Published online 2014 Mar 28. doi: 10.1186/1472-6882-14-112, PMCID: PMC3973604, PMID: 24678980

Yeh CW, Tseng LH, Yang CH, Hwang CF., (2019): Effects of oral zinc supplementation on patients with noise-induced hearing loss associated tinnitus: A clinical trial. Biomed J. 2019 Feb;42(1): 46-52. doi: 10.1016/j.bj.2018.10.009. Epub 2019 Mar 20.

CHAPTER 11

Nervous System, Musculoskeletal, Inflammation, Pain

The nervous system consists of the central nervous system made up of the brain and spinal cord and the peripheral nervous system made up of a network of nerves that ramify throughout the body. The musculoskeletal system made of bones and interactive muscles is also intimately connected to the nervous system. Pain resulting from musculoskeletal issues such as inflammation, arthritis involves the muscle ligaments, tendons, and inflammation of joints.

Many medical conditions affect both central and peripheral nervous systems. The causes and treatment for many of these conditions are still to be understood fully. The major conditions are:

1. Alzheimer's Disease

Alzheimer's disease affects brain cells and neurotransmitters affecting brain functions including memory. Brain scans show accumulation of amyloid and Tau proteins which is also seen in some patients having sleep apnea issues. It is also the most common form of dementia. Dementia is a syndrome associated with an ongoing decline in mental abilities. Age, lifestyle, heart health and family history are some of the factors that lead to this condition. Symptoms include forgetfulness, speech impediments, cognition, mood changes, and difficulty in performing normal functions. Treatment at present is based on a holistic approach involving drugs, electro-therapy, food, speech and exercise for body and brain. Herbalists consider that a range of herbal concoctions and certain types of herbal oil treatments also help in controlling onset or progress of this condition as well as other types of dementias.

There is evidence to suggest that single herbs or herbal formulations may offer certain complementary cognitive benefits to the approved drugs. Some of these herbals are:

Ginseng/Korean Ginseng/Chinese Ginseng (*Panax ginseng*): Kingdom: Plantae, Clade: Angiospermae, Eudicots, Order: Apiales, Family: Araliaceae.

Ginseng plants are perennial plants whose roots have medicinal properties. The main active ingredient is **panax saponin**, which can enhance psychomotor and cognitive performance, and can help Alzheimer's disease (AD) by improving brain cholinergic function and repairing damaged neuronal networks. However, the evidence for ginseng as a treatment of AD is inconclusive.

Common Sage (*Salvia officinalis*) Kingdom: Plantae, Clade: Angiospermae, Eudicots, Order: Lamiales, Family: Lamiaceae.

Salvia officinalis has been used in Greek herbal medicine as a diuretic, local anesthetic and for improving brain activity for many centuries. However, controlled clinical trials are inconclusive. **Toxicity:** The mono terpene thujone may be neurotoxic.

Two herbs gaining renown for cognitive support are Ashwagandha (*Withania somnifera*) and Brahmi (*Bacopa monnieri*).

Ashwagandha (*Withania somnifera*): Kingdom: Plantae, Clade: Angiospermae, Eudicots, Order: Solanales, Family: Solanaceae.

The roots of this perennial shrub growing in India hold the compounds **Withanolide A, withanoside IV, and withanoside VI**. These Compounds in Ashwagandha can promote construction of new neuronal networks. In a study, Oral administration of withanoside IV significantly improved memory deficits and prevented loss of axons, dendrites, and synapses in mice. In addition, **sominone**, an aglycone and a main metabolite of withanoside IV, significantly induced axonal and dendritic regeneration and synaptic reconstruction in damaged cultured rat cortical neurons.

Brahmi/Indian Pennywort (*Bacopa monnieri*) Kingdom: Plantae, Clade Angiospermae, Eudicots, Order: Lamiales, Family: Plantaginaceae.

Studies in rats fed with whole plant extracts of Brahmi showed statistically significant improvement in verbal learning, memory acquisition, and delayed recall in the Bacopa-treated group compared with controls.

Panchakarma (for Musculoskeletal- Arthritis as well as Alzheimer's, Parkinson's and other neural and neuro muscular conditions).

The traditional Indian Ayurvedic system also recommends Panchakarma treatment that involves diet, detoxifications using herbal laxatives and herbal oil treatments for preventing and or treatment of all Musculoskeletal diseases including **Arthritis** as well as Alzheimer's, **Parkinson's**, and other neural conditions. The treatment involves five steps namely, Basti= Herbalized oil enemas, Nasya= Nasal irrigation, Vamana= Therapeutic Vomiting, Virechana= Purgation, Raktamokshana= Blood Letting. Bloodletting and vomiting are no longer used and have been replaced by therapeutic herbal oil treatments. The aim of Panchakarma is to detoxify the body and clear the mind. The result as per Ayurvedic studies is that all toxins from body are cleared, the digestive system is healed, immunity is enhanced, stresses are removed, and body and mind are relaxed.

2. Parkinson's Disease

Parkinson's disease is a major neuronal disease affecting many in the older age group. It results from damage to the nerve cells in a region of the brain that produces dopamine, a chemical that is vital for the smooth control of muscles and movement. Symptoms include tremors, muscle rigidity or stiffness, slowing of movement, stooped posture, and balance problems. Parkinson's can also cause pain, depression and problems with memory and sleep.

Certain herbal treatments as single herbs or mixtures are said to control progress of Parkinson's. These herbs are:

Velvet Bean/Florida Velvet Bean/Bengal Velvet Bean (*Mucuna pruriens*), Kingdom: Plantae, Clade: Angiospermae, Eudicots, Order: Fabales, Family: Fabaceae (Leguminosae)

This annual climbing legume produces pinkish brown fruits in clusters. The fruits are covered with hair having an itch causing protein known as

Mucunain. The seeds are prescribed in the Ayurvedic and Unani systems of herbal medicine as a treatment for dementia, Alzheimer's disease, Parkinson's disease as well as for treating diabetes, sexual dysfunctions, inflammation, and snake bites. *M. pruriens* has been shown to have anti-Parkinson and neuroprotective effects. Mucuna spp. has the toxic compounds **L-dopa** and hallucinogenic tryptamines, and anti-nutritional factors such as phenols and tannins. Velvet bean is a commercial source of L-Dopa which is used in the treatment of Parkinson's disease. The seed extracts of this plant have natural **levodopa** as well as other chemicals that have neuroprotective effects. Levodopa is converted to dopamine in the brain. Some studies show that L-dopa derived from *M. pruriens* has many advantages over synthetic L-dopa when administered to Parkinson's patients, as synthetic L-dopa can have several side effects.

Ashwagandha/Indian Ginseng/Poison gooseberry/winter cherry (*Withania somnifera*): Kingdom: Plantae, Clade: Angiospermae, Eudicots, Order: Solanales, Family: Solanaceae.

These are shrubby tomentose perennials. As in the case of Alzheimers, ashwagandha is also considered a good herbal for treating Parkinson's disease. The plant has phytochemicals known as Withanolide

Turmeric (*Curcuma longa*): Kingdom: Plantae, Clade: Angiospermae, Monocots, Order: Zingiberales, Family: Zingiberaceae.

Ayurvedic practitioners prescribe curcumin-containing turmeric as an ingredient in medicines for treating Parkinson's disease and inflammatory conditions (Arthritis). However, **Curcumin** is not absorbed well into the blood stream and hence it is not clear how this can help in treating Parkinson's unless the drug delivery system includes black pepper extracts or liposomes. **Piperene** in black pepper is believed to enhance bioavailability of curcumin from turmeric.

Gou teng/cat's claw (*Uncaria rhynchophylla*): Kingdom: Plantae, Clade: Angiospermae, Eudicots, Order: Gentianales, Family: Rubiaceae.

This Chinese/Japanese traditional medicine herbal is used in treatment of hypertension, convulsive disorders (epilepsy), and for various head ailments such as headache or dizziness. It is reputed to have anti-spasmodic properties and has been used in China and Japan for more than 2,000 years to treat "the shakes," a common symptom of Parkinson's disease.

Cannabis/Marijuana/Hemp (*Cannabis sativa*) Kingdom: Plantae, Clade: Angiospermae, Eudicots, Order: Rosales, Family: Cannabaceae. Medicinal for multiple diseases.

Cannabis more popularly known as Marijuana or weed is one of the most well studied herbals at present because of its potential as a medicinal for a number of conditions including Cancers, pain, insomnia, Posttraumatic Stress syndrome (PTSD), neurodegenerative diseases, arthritis, inflammation, appetite control and several others. Taxonomically, *Cannabis sativa s*, *C. Sativa ssp indica* and *C. Sativa ssp. ruderalis* are three sub species, which make up the species Cannabis sativa. Those varieties with less than 0.3% THC are classified as Industrial hemp and is grown for seed oil and for fiber. Morphologically, the type species *C. sativa* ssp sativa is a tall plant with narrow leaves and less dense branches whereas, *C. Sativa* ssp.*indica* is a smaller but dense-bushier plant and C. Sativa ssp.ruderalis is a much smaller plant with thin fibrous leaves. All three-sub species and their strains have specific terpene compounds known collectively as **cannabinoids** as well several other phytochemicals. Of the 500+ chemicals found in Cannabis, 113 are Cannabinoids. The chief cannabinoids are 1. Tetrahydro cannabinol (THC), 2. Cannabidiol (CBD), 3. Cannabinol (CBN), which is a degradation product of THC, 3. Cannabigerol (CBG), 4. Tetrahydrocannabivarin (THCV), 5. Cannabidivarin (CBDV), 6. Cannabichromene (CBC). Of these, THC and CBD are the major bioactive components while the others also have important pharmacological functions.

Tetra hydro cannabinol ($C_{21}H_{30}O_2$) is the major psychoactive compound while CBD ($C_{21}H_{30}O_2$) is non-psychotropic. Although Cannabis resin and leaves have been in use for recreational and medical purposes for several centuries, interest in the Cannabinoids as a very powerful medicinal tool peaked after the discovery of the endo-cannabinoid system in humans and animals and the discovery of specific receptors known as CB_1 and CB_2 that bind to the endocannabinoids. It is now known that the plant cannabinoids referred to as Phyto cannabinoids also bind to these receptors. Thus, THC binds to CB_1 receptors found in the brain that accounts for its psychoactive powers. CB_2 receptors are found in the immune system particularly the spleen but the receptor has been detected in other tissues including brain. **CBD** does not bind directly to the CB_2 receptors but modulates this system indirectly.

From the above and from certain animal studies it appears that both THC and CBD have a role in modulating the functioning of the Central and peripheral nervous systems. Human studies on THC and CBD has been hampered because of regulatory reasons. Nevertheless, some clinical research on the non-psychoactive CBD has showed its usefulness in treating movement disorders including Parkinson's, epilepsy, anxiety, and pain. The drug Epidiolex based on CBD has been approved for treating certain types of epilepsy. Another drug Dronabinol/Marinol (synthetic THC) has been approved for treating patients having severe nausea and vomiting due to chemotherapy.

Many herbalists and even allopathic practitioners believe that there is great potential for Cannabis as a miracle plant for treating many medical conditions and that more research will reveal the extent and broad-spectrum nature of the use of Cannabis in treating diseases. Rather than single chemicals from the plant or synthetics of THC and CBD, the full benefits are likely to show up when whole plant having the many phytochemicals in the plant extracts are evaluated under clinical conditions. **Caution:** Although Cannabis especially those with high CBD are being considered as alternative treatment for Parkinson's disease, its use as a substitute for approved treatment options is not considered right at present.

3. Multiple Sclerosis (MS)

MS is a disease of unknown cause that manifests as multiple hard plaques of degeneration of the insulating layer of nerve fibers in the central nervous system. The loss of insulation allows, "Short circuiting" of nerve impulses. Depending upon where the degeneration occurs, patients may suffer paralysis, sensory disturbances, or blindness.

Ginger, Dandelion tea, Chamomile tea, Valerian root extracts, St. John's Wort, Echinaceae, Ginkgo, Kava Kava and stinging nettle have been suggested as potential herbals to help improve symptoms.

4. Epilepsy

Epilepsy is a specific condition that may occur at any age, seizures are more intense, longer lasting in duration, and recur with some frequency. A seizure is caused by disrupted electrical activity in the brain and the effect varies depending on the part of the brain involved. They can cause problems such as a loss of consciousness, unusual jerking movements (convulsions) as well as other unusual feelings, sensations, and behavior.

There are two main types of seizures namely generalized tonic-clonic seizures (grand mal) seizures and generalized absence seizures (Petit mal). Grand Mal are long lasting high frequency seizures that involve loss of consciousness, jerking and frothing in the mouth. Petit mal seizures usually start in childhood but can occur in adults. These seizures are brief and characterized by staring, loss of expression, unresponsiveness and stopping activity. Sometimes eye blinking or upward eye movements are seen. The person usually recovers at once and resumes their earlier activity, with no memory of the event.

Herbal Treatment

The FDA has approved the use of Epidiolex (a plant-based formulation of CBD from *Cannabis*) to treat seizures for people 2 years of age and older with Dravet syndrome and Lennox-Gastaut syndrome (LGS) which are seizure diseases. Extracts of Cannabis resin have been in use historically for treating Epilepsy. One of the main problems in using cannabis plant extracts rather than the synthetic CBD is the fact that there is considerable variation in the cannabinoid profile of different strains of Cannabis. Further anecdotal reports suggest that there is variation in the effectiveness of different strains for different individuals. The strain Charlotte's Web is easily the most famous cannabis strain for the treatment of seizures. Named after Charlotte, a young child stricken with epilepsy who inspired the development of this strain, it contains an average of 0.3% THC and contains high amounts of CBD, making it perfect for treating epilepsy without getting psychedelic. Charlotte's web is popularly used for children because of this quality. Charlotte's Web is recommended specifically for epilepsy only.

Cannabis web sites and pharmacies recommend the following strains for control of Seizures: 1. Indica varieties: Argyle, Athabasca, Bell ringer, Big Bang, Black Bubba, Brandywine, Darkside, Devils fruit, Fin Lou Zer Gorilla Biscuit, Grape Ape 2. Sativa varieties: Lemon Pie and Southern Lights,3. Hybrids: Aloha berry, Blackberry dream, Bedford Glue, Dreamers Glass, Frisian dew, Frosted Freak, Green Crack, GI001, G.O./A.T., Marcosus Marshmallow, Maui Haole, Red Dragon, White Widow,

5. Other Major Neurological Conditions

The following conditions may arise due to inherited genes or infection or environmental causes. There are no specific herbal formulations for treatment. However, in addition to allopathic medications, exercise, acupuncture etc, herbal massages may be of help in improving the symptoms.

1. Attention deficit/hyperactivity disorder (ADHD) 2. Motor neuron disease (MND) also known as Lou Gehrig's disease, amyotrophic lateral sclerosis, or ALS, 3. Neurofibromatosis 4. Sciatica 5. Shingles 6. Strokes 7. Guillain-Barre (gee-YAH-buh-RAY) syndromes 8. Ataxia 9. Autism 10. Catalepsy 11. Depression 12. Bipolar disease 13. Encephalitis 14. Locked-in Syndrome, 15. Meningitis 16. Huntingdon's disease 17. Tourette's syndrome, 18. Optic neuritis 19. Neuromyelitis Optica (Devic's disease) 20. Adrenoleukodystrophy and adrenomyeloneuropathy 21. Various inflammatory conditions including Rheumatoid arthritis.

Inflammation and Pain

The non-governmental organization "International Association for the Study of Pain (IASP) defines pain as "an unpleasant sensory and emotional experience associated with actual or potential tissue damage or described in terms of such damage". Pain is a response of the body towards a harmful stimulus such as a cut, injury, inflammation due to any reason including rheumatoid arthritis or other disease. Pain can be acute such as headache, injury, wounds etc when it is quick reaction to a stimulus and can resolve soon. Chronic pain on the other hand has a deeper underlying cause such as arthritis, cancer, peripheral neuropathy, various inflammatory conditions, or some unknown cause (idiopathic). Pain control medications may be an anesthetic that controls localized pain or systemic pain during and after surgery or they may be analgesics that can be used to control both acute and chronic pain. All anesthetics except cocaine are synthetic or semisynthetic while analgesics include opium derivatives such as morphine from the resin of Papaver somniferum, medical cannabinoids (CBD) from Cannabis sativa and salicylates from Salix. There are many other herbs and herbal formulations that have analgesic properties.

Anesthetics

Anesthetics are chemicals including herbal extracts that make patients insensitive to pain due to inflammation, musculoskeletal and other conditions so that surgery can be performed under anesthesia. Most anesthetics are synthetic non herbals and are either gases or injectables.

Coca (*Erythroxylum coca*): Kingdom: Plantae, Clade: angiospermae, Eudicots, Order: Malphigiales, Family: Erythroxylaceae

Coca plants are bushy plants that grow to a height of about 7-10 ft with tapering thin leaves. The plants are natives of South America and are grown legally and illegally in Columbia and other South American countries. It is the source of cocaine and the leaves, bark and other plant parts are chewed by the tribes of South and Central America as a stimulant and for control of pain. The powerful habit-forming hallucinogenic pain killer cocaine is obtained by processing the leaves and other parts of the plant. Modified cocaine known as Lidocaine is a local anesthetic.

Analgesics

These are agents that reduce pain involving the musculoskeletal system including various inflammations, rheumatoid and osteo arthritis and neural systems and include agents that are used post operatively or whenever there is pain.

Cannabis/Marijuana (*Cannabis sativa* ssp. *sativa*; *C. sativa* ssp. *indica*, *C. Sativa* ssp. *Ruderalis*): Kingdom: Plantae, Clade: angiospermae, Eudicots, Order: Rosales, Family: Cannabaceae.

Cannabis extracts and the CBD oil are being prescribed by Marijuana dispensaries for treating pain due to a variety of causes. The details of the plant and the cannabinoids associated with this plant have been described earlier.

Camellia (*Camellia sinensis*): Kingdom: Plantae, Clade: Angiospermae, Eudicots, Order: Ericales, Family: Theaceae. Extracts and infusions of Camellia are used in Traditional Chinese Medicine (TCM) for aches and pains, digestion, depression, detoxification and as an energizer to prolong life.

Camphor tree (*Cinnamomum camphora*) Kingdom: Plantae, Clade: angiospermae, Eudicots, Order: Laurales, Family: Lauraceae.

Camphor from the camphor tree is used in topical pain medications usually as a part along with salicylates, menthol, or other essential oils. Camphor is also a part of some anti-fungal creams and liquids. **Toxicity/Safety**: Camphor is FDA-approved for topical use as an analgesic and anesthetic in concentrations of 3% to 11%. Absorption of large quantities of camphor can cause severe reactions such as nausea, vomiting and even death.

The camphor tree is an evergreen tree belonging to the laurel family and is a native of China south of the Yangtze River, Taiwan, southern Japan, Korea, and Vietnam. It is an introduced species in many other countries. It grows up to 20–30 m (66–98 ft) tall.

Frankincense/Boswellia/Indian Frankincense/Shallaki/Salai Guggul/ Olibanum (*Boswellia serrata*): Kingdom: Plantae, Clade, Angiospermae, Eudicots, Order, Sapindales, Family: Burseraceae. Indian Frankincense is a medium sized deciduous tree with a light, spreading crown and drooping branches. It usually grows from 9 - 15 meters tall. The tree grows in Indian subcontinent and in North Africa. The bark, leaf powders as well as the resin and resin essential oil are used for medicinal purposes. The resin powder is encapsulated for internal consumption and resin as well as resin oil are mixed with other ingredients for topical applications for control of inflammatory conditions including osteo arthritis. An alcoholic extract of the root has shown anti-cancer activity against human epidermal carcinoma of the nasopharynx but there is little published data to support anti-cancer activity.

Frankincense/Boswellia/Olibanum/al-bakhūr/uunsii (*Boswellia sacra*): Kingdom: Plantae, Clade, Angiospermae, Eudicots, Order, Sapindales, Family: Burseraceae. The plants are found growing in the Arabian Peninsula and in the nearby middle east- Oman. Yemen, Somalia. The name is derived from its use by the Franks crusaders who brought it to the attention of Europe. The resin is commonly used in Eastern prthodox churches where the resin is burnt in rituals. The resin, essential oil, bark, root, and stem powders are considered to have medicinal properties including anti-inflammatory properties, cosmetics, and perfumery.

Myrrha/African Myrrha/African myrrh, herabol myrrh, Somali myrrhor, common myrrh, or gum myrrh (*Commiphora Myrrha*), Kingdom: Plantae, Clade Angiospermae, Eudicots, Order, Sapindales, Family, Burseraceae.: This is a scrubby spiny plant growing in the Arabian peninsula and in North and East Africa. The plant produces a resin called Myrrha from which an essential oil is extracted. The resin from this Arabian Peninsula and North Africa tree is used as an incense with disinfection properties. The resin and resin oil are part of both Ayurvedic, Unani/Arabian and Chinese herbal traditional medicine. Myrrha is used in mixtures with Frankincense to treat pain due to inflammation. It is also considered to be an emmenagogue and improve circulation. It is also a part of many different anti septic mouthwashes.

Opium poppy/Morphine/Opiates (*Papaver somniferum*): Kingdom: Plantae, Clade: angiospermae, Eudicots, Order: Ranunculales, Family: Papaveraceae.

Opium poppy plants are herbaceous annuals that produce strikingly beautiful flowers and exude a latex-resin upon wounding. The latex from the stem and from the fruit capsules is dried to generate opium.

The powerful analgesic opiate **Morphine** and alkaloids thebaine and oripavine are purified from the opium poppy resin. While morphine is used directly in the form of injections to control pain (mostly post-operative), the other two are used as bases to produce the semi-synthetic **oxycodone** and **hydrocodone** which are also powerful opiate analgesics. In addition, Opium also has another alkaloid namely codeine, which is a part of many, cough suppressants. All opiates are listed controlled substances and are available only by prescription. However, opium itself is added in many herbal formulations for inducing sleep and controlling pain but the growing of opium is controlled by Governmental agencies although illegal growing of opium poppy continue in many countries, particularly Afghanistan.

Salix/Willow (*Salix* sp.): Kingdom: Plantae, Clade: angiospermae, Eudicots, Order: Malphigiales, Family: Erythroxylaceae.

The genus Salix has over 350 different species ranging from creeping dwarf willows, short bushy shrubs to tall trees. All willows grow in temperate regions and even in the Arctic tundra. The bark exudes **Salicin**, which is converted to salicylic acid, that acts as an analgesic. The popular analgesic acetyl salicylic acid (aspirin) is derived from this. The discovery of salicylic acid from willows was the basis for the development of many modern non-steroidal anti-inflammatory drugs (NSAIDS). Herbal medicine practitioners steeped the bark of the willow to generate a decoction that was/is used as a pain killing medicine for headaches and other pain.

Thunder Vine/lei gong teng/, Radix (*Tripterygii wilfordii*) Kingdom: Plantae, Clade: Angiospermae, Eudicots, Order: Celastrales, Family: celastraceae.

Thunder vine extracts is used in TCM to treat arthritis, relieve pain, and reduce joint swelling. It can be extremely toxic. Both Chinese and United Kingdom regulatory agencies have issued toxicity warnings about the use of this herbal. One of the components of this tree namely **Triptolide** has shown both *in vitro* and *in vivo* activities against pancreatic cancer in mouse models but, the extreme toxicity of Triptolide has prevented further tests in this regard.

Other Herbs

The following plants have also been investigated for their analgesic properties based on folk medicine: *Margaritaria discoidea* (Phyllanthaceae); *Bridelia micrantha* (Phyllanthaceae), *Mikania natalensis* (Asteraceae), *Salvia africana-lutea:* (Lamiaceae); *Solenostemma sp:* (Apocyanaceae), *Detarium senegalense* (Caesalpineaceae), *Pycnobotrya* (Apocyanaceae), *Microloma saggittatum* (Apocyanaceae), *Motandra guineensis* (Apocyanaceae), *Salix mucronata* (Salicaceae), *Sceletium tortuosum* (Aizoaceae), *Strophanthus sarmentosus:* (Apocyanaceae), *Alphitonia petriei* (Rhamnaceae). Dry Ginger, Turmeric (curcumin), Ashwagandha, Guggul, and Giloy (Tinospora cordifolia) are other herbs that are reported to have a role in controlling pain through effects on inflammation and other properties.

Herbal Oil Bath

In the Indian subcontinent as well as in the middle east, it is a common practice to self-apply herbal-medicated oil (coconut or Sesame or Mustard) on the scalp and body followed by a rigorous self-massage and hot water bath. This is done two or three times a week during regular bath. This is different from Abhyanga massage where the oil is applied and massaged by trained professionals. This treatment helps relieve stress, musculoskeletal pain, cleanse the skin, and induce sleep and general wellbeing. The herbs used to prepare the medicated oils generally contain extracts of one or more of the following: Amla/Indian gooseberry/Nellikkai (*Phyllanthus emblica*), Ashoka (*Saraca indica*), Ashwagandha (*Withania somnifera*), Coco grass/Java grass/Muthanga/Mustak (*Cyperus rotundus*), Brahmi (*Bacopa Monera*), Camphor (*Cinnamomum camphora*), Holy Basil (*Ocimum sanctum*), Shellaki (*Boswellia*), Guggul, Shatavari *(Asparagus racemosus)*, Vetiver (*Vetiveria zizanioides*), Willow (*Salix* sp.), turmeric (*Curcuma longa*), Ginger (*Zingiber officianalis*), Saffron flowers (*Crocus* sp) and pepper (*Piper nigrum*).

Hallucinogens

Hallucinogens are agents that cause substantive changes in thoughts, consciousness, behavior, dreams, and other unreal experiences. These are psychoactive agents. All these are controlled substances because of their potential for abuse.

Dimethyltryptamine (DMT)

DMT, or dimethyltryptamine, is a hallucinogenic found in some plants as well as inside the brains, blood, and urine of mammals. South American Shamans use plant extracts having DMT in the religious drink Ayahuasca that causes severe hallucinations and visions of God. Many plants have DMT. These include various species of *Acacia* (Fabiaceae), *Delosperma* (Aizoaceae), *Desmodium* (Fabaceae), *Lespedeza* (Fabaceae), Mimosa (few species in family Fabaceae). It is also found in *Tetradium ruticarpum (Rutaceae), Euodia leptococca, Pilocarpus organensis, (Rutaceae), Limonia elephantum,* Several species of *Psychotria* (Rubiaceae), *Dictyoloma incanescens (Rutaceae), Dutaillyea drupacea and D. oreophila (Rutaceae), Tetradium ruticarpum (Rutaceae), Eriogonum sp. (Polygonaceae), Pandanus sp (Pandanaceae)., Virola venosa,* (Myristicaceae), *Horsfieldia superba,* (Myristicaceae), *Iryanthera macrophylla and, I. ulei,* (Myristicaceae) *Osteophloem platyspermum* (Myristicaceae), *Diplopterys cabrerana* (Malpighiaceae), *Nectandramegapotamica* (Lauraceae), *Petalostylis and labicheoides, P. cassinoides,* (Fabaceae), *Erythrina flabelliformis (Fabaceae), Phyllodium pulchellum* (Fabaceae), *Vepris ampody and Zanthoxylum arborescens* (Rutaceae).

Salvia Divinorum: **Kingdom: Plantae, Clade: Angiosperms, Clade: Eudicots, Clade: Asterids, Order: Lamiales, Family: Lamiaceae.**

Salvia Divinorum is a plant with psychoactive properties that is native to Mexico and Central and South America. Also called Diviner's Sage, Magic Mint, and Sage of the Seers, this hallucinogen distorts time and causes a "flying" feeling. Mazatec shamans use *Salvia divinorum* leaves to ease visionary states of consciousness during spiritual **healing sessions.**

Psilocybin

Psilocybin is a chemical compound contained in some psychedelic mushrooms native to Mexico, Central America, and the US. They are ingested orally or brewed in tea to reduce the bitter flavor. Psilocybin can produce hallucinations, panic attacks, and psychosis if consumed in large doses. It is also known as Magic Mushrooms and Shrooms.

Mescaline (Lophophora williamsii):

Mescaline is a hallucinogenic compound that is the active ingredient in Peyote (*Lophophora williamsii*) a small, spineless cactus. Mescaline is also found in *Opuntia acanthocarpa, O. echinocarpa, O. cylindrica* and *O. basilaris*. It is traditionally used by Native Americans in spiritual rites central to the Native American Church. It is a Schedule I substance. Abuse of Mescaline can cause illusions and hallucinations. Peyote and Mescaline are also known as Buttons, Cactus, Mesc, and Peyoto.

Nuciferine and Aporphine: Nuciferine and Aporphine are alkaloids that induce sedation and function as mild hallucinogens. The blue lotus and the holy sacred lotus both have these chemicals. Both are also known to reduce lipids and induce insulin production.

Blue lotus or lily, *(Nymphaea caerulea)*: Kingdom: Plantae, Clade: Angiospermae, Eudicots, Order: Nymphaeales, Family: Nymphaaceae. Sacred Lotus *(Nelumbo nucifera)*, Kingdom: Plantae, Clade: Angiospermae, Eudicots, Order: Proteales, Family: Nelumbanaceae. Both *Nymphaea caerulea and Nelumbo nucifera* which have the alkaloids **nuciferine** and **aporphine** were classified together earlier in the same family Nymphaceae but now separated as per the APG system. Recent studies have shown *Nymphaea caerulea* to have psychedelic properties. They have been used as a sacrament in ancient Egypt and certain ancient South American cultures and as a holy flower for Hindus. The flower extracts of both plants are used in aromatherapy.

Other Hallucinogens

1. **Christmas candlestick (*Lagochilus inebrians*):** Kingdom: Plantae, Clade: Angiosperms, Eudicots, Order: Lamiales, Family: Lamiaceae. Lagochilin is the active sedative, hypotensive and hemostatic chemical of this plant.

2. **Dream herb (*Calea zacatechichi/Calea ternifolia*):** Kingdom: Plantae, Clade: Angiosperms, Eudicots, Order: Asterales, Family: Asteraceae. Used as an herbal remedy in Mexico, leaf extract tea and smoking cause a dreamy state. Although not a controlled substance internationally, it is either banned or growth of plant is controlled in some states in the USA.

3. **Epena (*Virola* sp.):** Kingdom: Plantae, Clade: Angiosperms, Eudicots, Clade: Magnoliids, Order: Magnoliales, Family: Myristicaceae. Extracts of this south American plant are used to produce hallucinogenic snuff powders.

4. **False marijuana (*Zornia latifolia*):** Kingdom: Plantae, Clade: Angiosperms, Eudicots, Order: Fabales, Family: Fabaceae. This plant is a hallucinogenic substitute for recreational marijuana.

5. *Galbulimima belgraveana*: Kingdom: Plantae, Clade: Angiosperms, Eudicots, Clade: Magnoliids, Order: Magnoliales, Family: Himantandraceae. Plant berries are regarded as hallucinogens.

6. **Hawaiian Baby Woodrose (*Argyreia nervosa*):** Kingdom: Plantae, Clade: Angiospermae, Eudicots, Order: Solanales, Family: Convolvulaceae.

   It is mentioned in Food of the Gods as "a hallucinogenic substitute for cannabis". It is nicknamed Maconha brava because locals use it as a cannabis substitute

7. **Jimson weed (*Datura stramonium*), Deadly Nightshade (*Atropa belladonna*) and Mandrake (*Mandragora officinarum*):** Kingdom: Plantae, Clade: Angiosperms, Eudicots, Order: Solanales, Family: Solanaceae. Both are plants that are stimulants.

8. **Khat or qat (*Catha edulis*):** Kingdom: Plantae, Clade: Angiosperms, Eudicots, Order: Celastrales, Family: Celastraceae. In Africa, the young leaves are chewed as stimulants to cause euphoria.

9. **Morning glory/Beach moonflower (*Ipomoea tricolor, I. violacea*):** Kingdom: Plantae, Clade: Angiosperms, Eudicots, Order: Solanales, Family: Convolvulaceae

    The grounded seeds are used to make an infusion, which is used to induce a numbing trance for the purpose of divination and diagnosis. It is also employed by the Chontal people as a medicinal herb against gastrointestinal disorders, and is used as an appetizer, cathartic anti-dysentery remedy, and as a fever-reducing agent. Its psychedelic properties do not become clear until the user is asleep. Seeds have D-lysergic acid amide, lysergol, and turbicoryn, other lysergic acid alkaloids. All are psychoactive and hallucinogenic.

10. **Silene (*Silene capensis*):** Kingdom: Plantae, Clade: Angiosperms, Eudicots, Order: Caryophyllales, Family: Caryophyllaceae Produces vivid dreams after smoking.

11. **Sweet bay magnolia (*Magnolia virginiana*):** Kingdom: Plantae, Clade: Angiosperms, Magnolids, Order: Magnoliales, Family: Magnoliaceae.

    The leaves or bark have been placed in cupped hands over the nose and inhaled as a mild hallucinogen.

12. **woolly morning glory (*Rivea corymbosa/Argyreia nervosa*):** Kingdom: Plantae, Clade: Angiosperms, Eudicots, Order: Solanales, Family: Convolvulaceae. the seed extracts of this native plant from India induces psychedelic experiences. The phytochemical hallucinogens, **ergometrine, lysergol, lysergic acid** and other alkaloids in these seeds contribute to its pharmacological effects.

13. **Yopo (*Anadenanthera peregrina*):** Kingdom: Plantae, Clade: Angiosperms, Eudicots, Clade: Order: Fabales, Family: Fabaceae. Flower extracts are known to cause delirious flight of thoughts, often lasting many days. It has been used in healing ceremonies and rituals for thousands of years in South America.

Sleep and Insomnia

Natural Herbs for Insomnia

California poppy (*Eschscholzia californica*): Kingdom: Plantae, Clade: Angiosperms, Eudicots, Order: Ranunculales, Family: Papaveraceae.

California poppy is the state Flower of California. Plants are annuals or perennials growing in California, Oregon, Washington, and other western states of USA. This natural herb is commonly used as an herbal insomnia remedy. California poppy can be found in many herbal sleep aids sold in the United States today. This natural herb can help promote sleep, relaxation, and ease mild anxiety. Because of this herb's mild sedative properties, it is also safe to give to children who have trouble sleeping.

Catnip (*Nepeta cataria*): Kingdom: Plantae, Clade: Angiosperms, Eudicots, Order: Lamiales, Family: Lamiaceae.

Catnip is known for its calming effect and it has substances that are like those found in valerian. It has a more pleasant taste than valerian.

Chamomile (*Matricaria recutita*): Kingdom: Plantae, Clade: Angiosperms, Eudicots, Order: Asterales, Family: Asteraceae.

Chamomile has been used for centuries as an herbal sleep aid. As an herb for insomnia, it is drunk as a tea. It has a mild soothing effect, which aids sleep and reduces restlessness. It also has a positive effect on digestion. Chamomile does not lead to dependency. It has not been shown to have any side effects. Individuals with allergies to plants such as ragweed or daisies may have a reaction to this herb.

Kava Kava (*Piper methysticum*): Kingdom: Plantae, Clade: Angiosperms, Magnolids, Order: Piperales, Family: Piperaceae.

Kava kava is often used as an herbal insomnia remedy. This herb is extremely popular throughout the South Seas=Polynesia, and now throughout the United States. When used as natural sleep aids, kava kava can impart a natural calm feeling, as it helps relax the body, as well as enhance dreaming. This herb for insomnia is also often recommended for chronic fatigue. Long-term use should be avoided due to the possibility of liver damage.

Passionflower/Maypops (*Passiflora incarnata*): Kingdom: Plantae, Clade: Angiosperms, Eudicots, Order: Malphigiales, Family: Passifloraceae.

Plants are vines with violet, pink flowers and green fruits. All the aerial parts of the plant are used as herbal medicines for various conditions. *P. incarnata* has flavonoids and alkaloids, with leaves having the greatest concentration of flavonoids. Passiflora plant extracts are prescribed as a sleep aid.

Skullcap (*Scutellaria lateriflora*): Kingdom: Plantae, Clade: Angiosperms, Eudicots, Order: Lamiales, Family: Lamiaceae.

This herb with blue flowers growing in marshy and wetlands was used by the Aztec Indians of South America as a sedative, has a calming, sleep inducing effect. This is a gentle, non-habit-forming herbal sleep aid. This is a very safe herb and has been used by both adults and children to counter the effects of stress and tension. It helps relax the mind and body to induce restful sleep. The active ingredient, **harmine**, and related compounds help inhibit the breakdown of serotonin. This herb can be taken as a tea, tincture or as capsules. Traditionally, it is used for restless leg syndrome and other causes of insomnia. Skullcap relieves nervous tension and renews the central nervous system.

Hops (*Humulus lupulus*): Kingdom: Plantae, Clade: Angiosperms, Eudicots, Order: Rosales, Family: Cannabaceae.

The cone shaped flowers are used to flavor beer and give it the bitter taste. Hops is a common flavoring for beer which also has a calming, sedative effect and it has been used for millennia to treat insomnia, anxiety, agitation and to relieve pain. The soothing substance found in hops is **methylbutanol**, which has a calming effect on the central nervous system. Hops can be consumed as herbal tea, or in tincture or capsules form that may also have other sleep-enhancing herbs like valerian. Pillows are sometime made of hops to help induce relaxation and restful sleep.

Valerian (*Valeriana officinalis*): Kingdom: Plantae, Clade: Angiosperms, Eudicots, Order: Rosales, Family: Cannabaceae.

Valerian root extract is the most popular herbal for insomnia, and it is the natural source of **Valium**. It eases nervous tension, muscle tension and anxiety. It can be used as an occasional treatment for restlessness but may be most beneficial for insomnia treatment over the long term. It works well in combination with other sedative herbs. Valerian root does not have the harsh side effects of many pharmaceutical treatments, like Valium. In high doses, this herb can cause nausea, headaches, dizziness, weakened heartbeat and even paralysis. Taken in recommended dosages, it is considered safe. It should not be taken when driving or during any actions requiring mental acuity. It should not be combined with pharmaceutical medicines that have similar effects.

Other Sleep Aid Herbs

Other natural herbs that have been used as herbal sleep aids include: Cowslip – (*Primula veris*), Great Mullein – (*Verbascum thapsus*), Mugwort – (Artemisia vulgari), Bugleweed – (*Lycopus europaeus*), Jamaica Dogwood – (*Piscidia erythrina*), Mistletoe – (*Viscum album*), Lemon Grass – (Cymbopogon citratus), Lavender – (Lavandula angustifolia), Vervain – (Verbena officinalis), Heather – (*Calluna vulgaris*), American Ginseng – (*Panax quinquefolius*), Common Mallow – (*Malva sylvestris*), Sweet Basil – (*Ocimum basilicum*).

References

https://worldherbals.com/en/zornia-latifolia-maconha-brava-p-153.

https://jnnp.bmj.com/content/75/12/1672.

Akhondzadeh, S, Noroozian, M, Mohammadi, M, Ohadinia, S, Jamshidi, AH and Khani M: Salvia officinalis extract in the treatment of patients with mild to moderate Alzheimer's disease: a double blind, randomized and placebo-controlled trial. J Clin Pharm Ther. 2003, 28: 53-59., 10.1046/j.1365-2710.2003. 00463.x

Bar-Lev Schleider L, Mechoulam R, Lederman V, Hilou M, Lencovsky O, Betzalel O Shbiro L, Novack V (2018): Prospective analysis of safety and efficacy of medical cannabis in large unselected population of patients with cancer. European Journal of Internal Medicine 2018 Mar. Vol. 49; 37-43

Birks J, Grimley Evans J: Ginkgo biloba for cognitive impairment and dementia. Cochrane Database Syst Rev. 2009, 1: CD003120-

Boggs, Douglas L; Nguyen, Jacques D; Morgenson, Daralyn; Taffe, Michael A; Ranganathan, Mohini (6 September 2017). "Clinical and preclinical evidence for functional interactions of cannabidiol and Δ9-tetrahydrocannabinol". Neuropsychopharmacology. 43 (1): 142–154. doi: 10.1038/npp.2017.209. ISSN 0893-133X. PMC 5719112. PMID 28875990.

Caroline Turcotte, Marie-Renée Blanchet, Michel Laviolette, and Nicolas Flamandcorre (2016): The CB2 receptor and its role as a regulator of inflammation. Cell Mol Life Sci. 2016, 73(23): 4449–4470, Published online 2016 Jul 11. doi: 10.1007/s00018-016-2300-4, PMCID: PMC5075023, PMID: 27402121.

Dhopeshwarkar A, Mackie K (2014): CB2 Cannabinoid receptors as a therapeutic target-what does the future hold? Mol Pharmacol. 2014 Oct,86(4): 430-7. doi: 10.1124/mol.114.094649. Epub 2014 Aug 8.

Fu LM, Li JT: A systematic review of single Chinese herbs for Alzheimer's disease treatment. Evid Based Complement Alternat Med. 2009,

Iwasaki K, Kobayashi S, Chimura Y, Taguchi M, Inoue K, Cho S, Akiba T, Arai H, Cyong JC, Sasaki H: A randomized, double-blind, placebo-controlled clinical trial of the Chinese herbal medicine 'Ba wei di huang wan' in the treatment of dementia. J Am Geriatr Soc. 2004, 52: 1518-1521. 10.1111/j.1532-5415.2004.52415. x.

Jayaprakasam, B, Padmanabhan, K, Nair, M.G. Withanamides in Withania somnifera fruit protect PC-12 cells from β-amyloid responsible for Alzheimer's disease. Phytother Res. 2010,24(6): 859-863.

Katzenschalger R, Evans A, Manson A, Patsalos PN, Watt H et al (2004): Mucuna Pruriens in Parkinson's Disease: A Double-blind Clinical and Pharmacological Study –Journal of Neurology, Neurosurgery and Psychiatry, 2004. 75: 2672-7.

Li J, Wu HM, Zhou RL, Liu GJ, Dong BR (2008): Huperzine A for Alzheimer's disease. Cochrane Database Syst Rev. 2008, 2: CD005592-

Lucia Raffaella Lampariello, Alessio Cortelazzo, Roberto Guerranti, Claudia Sticozzi, and Giuseppe Valacchi, (2012): The Magic Velvet Bean of Mucuna pruriens, J Tradit Complement Med. 2012 Oct-Dec; 2(4): 331–339., PMCID: PMC3942911, PMID: 24716148.

Mazza M, Capuano A, Bria P, Mazza S: Ginkgo biloba and donepezil: a comparison in the treatment of Alzheimer's dementia in a randomized placebo-controlled double-blind study. Eur J Neurol. 2006, 13: 981-985. 10.1111/j.1468-1331.2006.01409. x.

Miao YC: Part 5: a randomized, double blind and parallel control study of GEPT extract in the treatment of amnestic mild cognitive impairment. PhD thesis. 2008, Beijing University of Chinese Medicine, 95-117.

Pathak-Gandhi N, Vaidya, A. D. (2017): Management of Parkinson's disease in Ayurveda: Medicinal plants and adjuvant measures. J Ethnopharmacol. 2017 Feb 2,197: 46-51., doi: 10.1016/j.jep.2016.08.020. Epub 2016 Aug 17.

Rätsch, Christian (2005): The Encyclopedia of Psychoactive Plants: Ethnopharmacology and Its Applications. Inner Traditions/Bear.

Sathiyanarayanan L, Arulmozhi S. (2007): Mucuna pruriens A comprehensive review. Pharmacognosy Rev. 2007; 1: 157–162.

Spencer J.P.E, Jenner A, Butler J, Aruoma O.I, Dexter D.T, Jenner P, Halliwell B. Evaluation of the pro-oxidant and antioxidant actions of L-Dopa and dopamine in vitro: implications for Parkinson's disease. Free Rad. Res. 1996; 24: 95–105.

Snitz BE, O'Meara ES, Carlson MC, Arnold AM, Ives DG, Rapp SR, Saxton J, Lopez OL, Dunn LO, Sink KM, DeKosky ST, (2009): Ginkgo Evaluation of Memory Study I: Ginkgo biloba for preventing cognitive decline in older adults: a randomized trial. JAMA. 2009, 302: 2663-2670., 10.1001/jama.2009.1913.

Srivastav, S., Fatima, M. and, Mondal, A. C., (2017): Important medicinal herbs in Parkinson's disease pharmacotherapy. Biomed Pharmacother. 2017 Aug.92: 856-863., doi: 10.1016/j.biopha.2017.05.137.

Tian J, Shi J, Zhang L, Yin J, Hu Q, Xu Y, Sheng S, Wang P, Ren Y, Wang R, and Wang Y(2009): GEPT extract reduces a beta deposition by regulating the balance between production and degradation of amyloid-beta in APPV717I transgenic mice. Curr Alzheimer Res. 2009, 6: 118-131. 10.2174/156720509787602942.

CHAPTER 12

Anti-Parasitic and Anti-Microbial Herbs

Many diseases are caused by parasites. Other diseases are caused by viruses, bacteria, fungi, and mycoplasma. Many herbal concoctions are prescribed by herbalists for treating diseases caused by these agents. These are:

Artemesia/Wormwood (*Artemesia annua*): Kingdom: Plantae, Clade: Angiospermae, Eudicots, Order: Asterales, Family: Asteraceae.

Artemisia, which is commonly known as wormwood, has the powerful anti-malarial compound Artemesinin besides many **coumarins, flavones, flavanols** and many phenolic compounds. Of these, Artemesinin has been shown experimentally and clinically to be effective against the malarial parasite *Plasmodium falciparum*. Artemesinin-based drugs are prescribed for treating malaria. However, Artemesinin resistance has been reported from Thailand and Cambodia. The related species *A. cina* and *A. brevifolia* are the source of another anti-parasitic drug Santonin. Santonin paralyses the helminths and kills them. However, **Santonin** has been replaced by other more safe and powerful acaricides.

Acalypha (*Acalypha integrifolia*): Kingdom: Plantae, Clade: Angiospermae, Eudicots, Order: Malphigiales, Family: Euphorbiaceae.

Acalypha indica is a soft stemmed herb with catkin inflorescence and is listed in the Indian pharmacopeia. Extracts of leaf and stem are used as vermifuges and as a treatment for skin infections as per the Siddha and Ayurvedic systems of India. **Toxicity:** all species have cyanogenic glycosides and hence, poisonous if consumed in large quantities. Ingestion of herbal medicine having *Acalypha indica* may lead to hemolysis in patients suffering from glucose-6-phosphatase dehydrogenase deficiency. Plants are branching shrubs found growing wild in many parts of Africa. Leaf decoctions are drunk as a purgative to end intestinal worms. Leaf extracts are smeared to treat skin scabies infections. Several sub species of Acalypha grow in tropical islands and in the Indian sub-continent. The leaves, stems and roots have **saponins, tannins, sterols, terpenes,** and traces of alkaloids. The methanol extract of stem and leaves of this plant showed potent antibacterial activity against *Escherichia coli* (EC)), *Pseudomonas aeruginosa,* and *Staphylococcus aureus.* The highest antibacterial activity was seen against clinical isolates of *Enterococcus faecalis, S. aureus,* and Methicillin-Resistant *S. aureus* (MRSA.). Moderate antifungal activity was seen against *Candida albicans.*

African Custard Apple/Wild Custard Apple (*Annona senegalensis*) Clade: Magnolids, Order: Magnoliales, Family: Annonaceae.

This shrubby tree is a native of east and central Africa. It is closely related to the cultivated custard apple bearing sugary fleshy fruits. The bark as well as roots of the plant is extracted, and the aqueous decoction is prescribed as a tonic and medicine to kill and remove intestinal worms. The seeds have insecticidal properties and hence seed powders and extracts are used as organic insecticide to control insect infestation of crop plants. The related *Annona squamosa* is cultivated and has a powerful insecticide in its seeds. In India and Thailand, the leaf extracts of *A. squamosa* are used to control dysentery.

***Strophanthus welwitschii*: Kingdom: Plantae, Division: Angiospermae, Class: Eudicots, Order: Gentianales, Family: Apocynaceae.**

Strophanthus welwitschii grows as a deciduous shrub or small tree up to 5 meters (16 ft) tall, or as a liana up to 8 meters (26 ft) long, with a stem diameter up to 10 centimeters (4 in). Its fragrant flowers feature a white turning purple corolla, creamy and red or purple-streaked on the inside. Its habitat is forests or rocky woodlands from 300 meters (1,000 ft) to 1,800 meters (6,000 ft) altitude. *S. welwitschii* is used in local medicinal treatments for respiratory conditions, gonorrhea, and scabies. The plant has been used as arrow poison. *Strophanthus welwitschii* is native to Democratic Republic of Congo, Tanzania, Angola, and Zambia.

Common cabbage Tree (*Cussonia spicata*): Kingdom: Plantae, Clade: Angiospermae, Eudicots, Order: Apiales, Family: Araliaceae.

This is a tree growing in sub Saharan Africa. The decoction made from the bark of this tree has anti-malarial properties. It is bitter to taste and can be added to tonic water as a substitute for quinine.

Garlic (*Allium sativum*): Kingdom: Plantae, Clade: Angiospermae, Monocots, Order: Aspergales, Family: Amaryllidaceae.

Garlic bulb extracts are known to have anti-microbial and anti-nematode activities. It has several organo-sulfur compounds that are responsible for the acrid smell and anti-microbial activities of garlic. Plants are fleshy bulbous herbs. The stem is a bulb growing underground.

In India, it is an important component of certain cough syrups and formulations for respiratory conditions.

Ipecac (*Carapichea ipecacuanha*): Kingdom: Plantae, Division: Angiospermae, Class: Eudicots, Order: Gentianales, Family: Rubiaceae. The bark and leaf extracts have anti-helminthic properties and are used in African native medicine to treat intestinal worm infections.

The plants are perennial weeds that grow to a height of about 4 ft. They produce small green flowers in panicles. Seeds are small and have an essential oil rich in the anti-worm **"Ascaridole" terpene. Health:** Excess doses of the oil or extracts can cause severe dizziness, vomiting and even death. The leaves have oxalates and hence culinary usage should be limited. The red sap of the plant is used to treat scabies and tapeworm. Bark Decoction is used to treat malaria and jaundice. This latex is used to treat scabies and tapeworm infestations. The roots of this South American plant have the alkaloid **emetine**, which is a powerful emetic and **anti-amoebic** agent. The plant extracts cause severe vomiting and nausea, but the purified alkaloid can be administered intra muscularly and is active against amoebae. Root extracts are also components of the Ipecac cough syrup. **Toxicity:** The use of Emetine has been stopped for treating amoebas. Emetine in small doses is also an expectorant and is useful to induce vomiting in cases of drug overdoses.

Male fern (*Dryopteris filixmas*): Kingdom: Plantae, Clade: Polypodiopsida, Order: Polypodiales, Family: Polypodiaceae.

This is a common fern found growing in Europe and other temperate climes. Root extracts are used to expel tapeworms but there are more powerful and safer anti-tapeworm medications. The liquid extract is one of the best anthelmintic against tapeworm, which it kills and expels. It is usual to administer this worm medicine last thing at night after several hours of fasting and to give a purgative such as castor oil first thing in the morning. A single sufficient dose will often cure at once. The drug is employed for similar purposes by veterinary practitioners in the powdered form. In too large doses, however it is an irritant poison causing muscular weakness and coma.

Neem (*Azadirachta indica/Melia azadirachta*) Kingdom: Plantae, Clade: Angiospermae, Eudicots, Order: Sapindales, Family: Meliaceae.

Neem leaves, flowers, fruits, and seeds have powerful anthelmintic, antifungal, antibacterial, and antiviral activities. Extracts from this plant have been in use since ancient times in India as a medicinal plant with above-mentioned medicinal properties. The plant is considered holy in India and is worshipped as such mainly because it was considered as being active against the smallpox disease.

Neem plants are large evergreen trees found in the Indian subcontinent but now it is grown extensively in Mexico, some parts of N and South America and in Africa and Asia. The trees are highly branched with compound serrated glabrous leaves. Flowers are borne in inflorescences and are small and white in color. Fruits are elongated globules and are green becoming yellow green upon maturity. The fruits have a single seed. Neem seed oil is commercially used as a bio insecticide treatment to control insect infestation of crops and ornamental plants.

Quinine tree/Jesuits bark (*Cinchona officianalis*): Kingdom: Plantae, Clade: Angiospermae, Eudicots, Order: Gentianales, Family: Rubiaceae

Quinine trees are small evergreen trees with ridged bark and ovate opposite simple glabrous leaves. Plants are natives of South America and have been introduced elsewhere especially Indonesia, India, China, and South Asia islands. The bark has quinine and quinidine. The bark is steeped in water to produce a very bitter decoction, which is then prescribed as a drink to control malaria. Bark is also made into tablets, oils, and other vehicles for consumption for medicinal purposes. The alkaloid quinine is extracted from the bark. Apart from malaria, quinine and cinchona plant extracts are also prescribed for treating neuralgia, various types of spasms and muscular cramps. It is also a major part of homeopathic pharmacopeia.

Stevia (*Stevia rebaudiana*): Kingdom: Plantae, Division: Angiospermae, Class: Eudicots, Order: Asterales, Family: Asteraceae. Stevia leaves have the artificial sweetener based on steviol glycosides. Plants are reported to have compounds that kill the tick-borne spirochete *Borrelia burgdorferi*. The plants are perennial herbs grown commercially for extracting no calorie sweetener.

Thyme (*Thymus vulgaris*): Kingdom: Plantae, Division: Angiospermae, Class: Eudicots, Order: Lamiales Family: Lamiaceae. Chromosome number, 2n = 24.

Thyme is a bushy evergreen shrub with small, highly aromatic, grey-green leaves and clusters of purple or pink flowers in early summer. Leaves have **Thymol** that is a disinfectant. Thymol is a powerful anti-bacterial and antifungal agent and is a part of many mouthwashes and disinfectants.

Wormseed (*Chenopodium ambrosioides anthelminticum/Dysphania ambrosioides*): Kingdom: Plantae, Division: Angiospermae, Class: Eudicots, Order: Caryophyllales, Family: Amaranthaceae/ Chenopodiaceae. Other names: Chenopodium anthelminticum, C. integrifolia, Atriplex ambrosioides.

Wormseed extracts/oil is used in central and south America to expel parasitic worms from the body as a vermifuge and is a good anti-amoebic agent. As in the case of other herbs, it has also been used traditionally for treating other ailments such as hemorrhoids, anti-asthmatic, and stomach pain.

Poly herbal formulation/Nilavembu Kudineer: Herbal constituents are: Bile of Earth/Kalamegh/Nilavembu (*Andrographis paniculata*),: Acanthaceae, ; Vettiver (*Vetiveria zizanioides*): Poaceae; Indian

Sandalwood (*Santalum album*): Santalaceae; Serpent gourd (Trichosanthes cucumerina) – Cucurbitaceae; Threadstem carpet weed/Parpatakam (*Mollugo cerviana*)- Molluginaceae; Ginger rhizome/Inji(*Zingiber officinale*)- Zingiberaceae, Black pepper corn (*Piper nigrum*)- Piperaceae, Nut grass/Coco grass/Korai Kilangu (*Cyperus rotandus*).

Nilavembu Kudineer is a polyherbal ethanol-aqueous formulation based on Siddha medicine having extracts of *Andrographis paniculata, Vetiveria zizanioides, Santalum album, Trichosanthes cucumerina, Mollugo cerviana, Zingiber officinale, Piper nigrum* and *Cyperus rotandus*. This oral formulation was found to improve and relieve symptomatic fevers, inflammation, arthritic pain, break-joint pain, and jaundice associated with infections of both Dengue virus as well as Chikungunya viruses. This polyherbal formulation is sold in India as a dry mix of these herbs in pouches and the herb mix is extracted with water before consumption during these viral infections.

Brazilian Iron wood/Pau Ferro/Leopard Tree (*Caesalpinia ferrea/ Libidibea fer*rea); Kingdom: Plantae, Division: Angiospermae, Class: Eudicots, Order: Fabales, Family: Fabaceae.

Extracts of this plant as well as those of *Brosimum acutifolium*, and *Salacia impressifolia* were evaluated for anti-fungal activity against three strains of *Candida albicans, C. glabrata* and *C. Guillermondii* grown in culture. Extracts of *C. ferrea* were extremely effective as cytotoxic and anti-fungal activity against all three strains of Candida.

The Brazilian Iron wood plant is a perennial tree growing to a height of 40 ft with bipinnate compound leaves and yellow flowers, borne in terminal or axillary inflorescences; The individual blooms are tiny and bell-shaped. Fruits are hard, flat, brown pod with several seeds. These plants grow naturally in Brazil and Bolivia. The wood of this plant is used for commercial purposes. There is evidence that exposure to the plant extracts, wood powder and pollen can cause severe allergies in susceptible individuals.

References

http://www.theplantlist.org/tpl1.1/record/gcc-39740

https://prota4u.org/database/protav8.
asp?g=pe&p=Acalypha+integrifolia+Willd.

https://www.prota4u.org/database/protav8.
asp?g=pe&p=Acalypha+indica+L.

https://pfaf.org/user/Plant.aspx?LatinName=Cinchona+officinalis

https://indiabiodiversity.org.

http://plantselector.botanicgardens.sa.gov.au.

https://www.sciencedirect.com/science/article/pii/S1876619615002429.

http://www.theplantlist.org/tpl/record/gcc-99074.

https://plants.jstor.org/stable/10.5555/al.ap.specimen.k001119018.

https://www.africamuseum.be/en/research/collections_libraries/biology/prelude/browse_plants.

Clarkson, C; Maharaj, V., Crouch, N., and Grace, O.M., (2004): i n vitro antiplasmodial activity of medicinal plants native to or naturalized in South Africa, Journal of Ethnopharmacology 92(2-3): 177-91 June 200.

Cletus Ukwubile (2012): Anti-Helminthic Properties of Some Nigerian Medicinal Plants on Selected Intestinal Worms in Children (Age 5-13) in Ogurugu, South East Nigeria, Journal of Bacteriology & Parasitology 03(09) January 2012, DOI: 10.4172/2155-9597.1000159.

Jain, J., Kumar, A., Narayanan, V., Ramaswamy, R.S., Sathiyarajeswaran, Shree Devi, P., M.S., Kannan, M., and Sunil, S. (2018) Antiviral activity of ethanolic extract of Nilavembu Kudineer against dengue and Chikungunya virus through in vitro evaluation, Journal of Ayurveda and Integrative Medicine, https: //doi.org/10.1016/j.jaim.2018.05.006.

Mônica Regina Pereira Senra Soares, César Augusto Caneschi, Maria das Graças Afonso Miranda Chaves, Marcela Mota, Pedro Henrique

Fazza Stroppa, Wagner Barbosa, Nádia Rezende Barbosa Raposo I(2018): in vitro antifungal activity and cytotoxicity screening of dry crude extracts from Brazilian Amazonia plants. African Journal of Traditional, Complementary and Alternative Medicines, Vol 15, No 4 (2018) 13-21.

Roumita Seebaluck-Sandoram, Lall, M., Fibrich, B., Blom van Staden, A., Fawzi Mahomoodally (2018): Antibiotic-potentiating activity, phytochemical profile, and cytotoxicity of Acalypha integrifolia Willd. Journal of Herbal Medicine, Volume 11, March 2018, Pages 53-59 (Euphorbiaceae).

Zafar Iqbala Muhammad Lateefa Muhammad Ashrafb Abdul Jabbara Anthelmintic activity of Artemisia brevifolia in sheep, Journal of EthnopharmacologyVolume 93, Issues 2–3, August 2004, Pages 265-268.

CHAPTER 13

Longevity, Ageing, and Herbals

Ageing is a normal inevitable process involving mental, physical, psychological, and social changes. The process of ageing results in greying and or loss of hair, wrinkled skin, speech impairment, cognitive problems, hearing and loss of sight, stumbling, falls as well as forgetfulness leading to dementia and even Alzheimers disease, Parkinson's disease, susceptibility to infections, diabetes, cancer, high blood pressure and other symptoms. Individuals are living longer due to improved hygiene, nutrition, and better healthcare options. Modern research is offering options for improvement in all aspects of the process with the aim of not only prolonging life but also, also ageing with grace. Extension of life as well as delaying the ageing process and associated symptoms is being tried by various experimental approaches. Thus, enhancement of autophagy, elimination of senescent cells from the body, transfusion of plasma from blood of young people, physical exercise, intermittent fasting, increasing adult neurogenesis, adding antioxidant fruits and vegetables to diet, intake of senolytic drugs, boosting telomerase activity and stem cell therapy are some of the approaches to delay onset of age-related diseases and symptoms. Several studies show that the administration of autophagy enhancers, senolytic drugs, plasma from new people, drugs that enhance neurogenesis and brain-derived neurotrophic factor (BDNF) are promising approaches to sustain normal health during aging. Stem cell therapy has also shown promise for improving regeneration and function of the brain. Many studies point to the importance of scavenging free radicals that destroy cells through proper nutrition and addition of flavanoids and antioxidants to the diet. In other studies, enhancement, or restoration of telomerase activity in cells of older individuals could be important.

One of the most studied proteins in the past ten years that has been found to aid the aging process is called **sirtuin**, or silent information regulator. Sirtuins control the rate at which we age, and the length of our lifespan, controlled by "longevity genes. It is thought that by stimulating these longevity genes, one can extend life span.

Apart from these approaches, several studies have shown that regular exercise and intermittent fasting can improve both qualities of life as well as extend life span. Many added studies on the intake of proper nutraceuticals and herbal formulations can delay onset of age-related diseases and even extend life.

Before listing and describing the role of herbs in the anti-ageing process, I will give a brief description of the various phenomena that are believed to be involved in ageing. In some unknown way, certain herbs and their preparations activate these processes and help delay the onset of ageing, improve quality of life, and enhance life span.

1. Autophagy

Autophagy is a Lysosomal disintegration process, which is a protective housekeeping machinery to get rid of damaged cell organelles, longstanding improperly folded proteins and colonizing pathogens. Autophagy shields cells against stress. Autophagy has a significant role in the modulation of the aging process. The normal process of autophagy slows down as we age and hence any process that can restore autophagy may slow-down aging. Thus, longevity increase achieved through food deprivation and calorie restrictions (CR) are helped through up-regulation of autophagy.

2. Apoptosis

This is a form of programmed cell death, involving coordinated and stepwise series of biochemical reactions resulting in the ordered disassembly of a cell from an organism. This normal biological process is needed for proper organ development during embryogenesis and for the removal of abnormal cells. Cancer cells are resistant to apoptosis. The process of aging increases the chances of cancer and hence a proper understanding of programmed cell death process is important. Consequently, manipulation of the apoptotic process is essential to understand the development of various diseases and to discover potential therapeutic targets.

3. Senescent Cell Elimination as an Anti-Aging Therapy, Senolytics

Cellular senescence (ageing cells) is typified by an irreversible cell-cycle arrest, a resistance to apoptosis, and acquisition of a pro-inflammatory, tissue-destructive senescence-associated secretory phenotype. Senescent cells cannot take part in standard tissue preservation and tissue repair because they do not proliferate. Senescence is a common feature during the aging process with an age-related increase in the number of senescent cells. The elimination of senescent cells using drugs known as **senolytics** would slow down aging and support better function during old age. Anti-senescence activity has been found in a variety of molecules, including synthetic senolytic drugs and natural compounds in certain types of food. Senolytic drugs are agents that selectively induce apoptosis (death) of senescent cells. These cells accumulate in many tissues with aging and at sites of disease in multiple chronic diseases. In studies in animals, targeting senescent cells using genetic or pharmacological approaches delays, prevents, or alleviates multiple age-related phenotypes, chronic diseases, geriatric syndromes, and loss of physiological resilience. Among the chronic conditions successfully treated by depleting senescent cells in preclinical studies are frailty, cardiac dysfunction, vascular hypo reactivity and calcification, diabetes mellitus, liver steatosis, osteoporosis, vertebral disk degeneration, pulmonary fibrosis, and radiation-induced damage. One of the first clinical studies involved the treatment of idiopathic pulmonary fibrosis (IPF) by administering a combination of two senolytics namely **dasatinib and quercetin** (DQ), taken by mouth for three consecutive days each week for three consecutive weeks (nine doses total). The most consistent improvements following senolytic therapy were seen in participants' mobility.

Transfusion of Plasma from Young Individuals to Promote Successful Aging

Transfusion of blood and blood letting are age-old therapies in the field of medicine. This procedure is being revived for limiting or reversing aspects of aging in various organs throughout the body by involving transfusion of blood or plasma from younger individuals to the aged. It is thought that molecules circulating in the new people can rejuvenate the aging cells and tissues. For example, experimentally, transfusion of blood from a young animal to an aged animal enhanced adult neurogenesis as well as synaptic plasticity in the aged brain.

Intermittent Fasting

Diet and eating schemes have a significant influence on the pathogenesis of many age-associated diseases. one of the most efficient nongenetic dietary interventions that can enhance longevity is calorific restriction (CR), involving reduced calorie intake by about 20-40% as much or as often as necessary or desired. A study on the dietary intake of adults living in Okinawa, a Japanese island having nearly five times higher number of centenarians than any other part of the world, revealed that Okinawan's consumed 17% fewer calories than the average adult in Japan and 40% less than the average adult in the United States. Many investigations in animal prototypes have proved that reduced food intake leads to increased lifespan.

Physical Exercise for Modulating Aging and Preventing Dementia

The benefits of regular physical exercise (PE) for conserving the function of the cardiovascular, musculoskeletal, and nervous systems are well documented. PE boosts blood flow to the working skeletal muscles by up to 100-fold and moderately to the brain, while organs such as liver, kidney, and testes meet diminished blood flow during PE. Even with the variability in blood flow to different organs, all organs gain from regular PE.

Oxidative Stress

Oxidative stress has a significant influence in the development of many age-related diseases, which include arthritis, diabetes, dementia, stroke, cancer, atherosclerosis, vascular diseases, obesity, osteoporosis, and metabolic syndromes. Older adults are more susceptible to oxidative stress due to a reduction in the efficiency of their endogenous antioxidant systems. Emerging research suggests that natural antioxidants can control the autoxidation by interrupting the propagation of free radicals or by inhibiting the formation of free radicals. Through such actions, antioxidants reduce oxidative stress, improve immune function, and increase healthy longevity. Therefore, dietary supplementation with antioxidants has received considerable interest in combating aging. This approach is also consistent with the free radical theory of aging that lowering the global level of oxidative stress in the body would retard aging, increase lifespan, and prevent and treat aging-associated diseases. However, clinical trials with antioxidants in the elderly population have reported inconsistent findings. The most promising antioxidants currently in clinical trials for treating cognitive dysfunction in aging and AD include resveratrol (RESV) and curcumin (CUR).

Stem Cell Therapy

The efficacy of intracerebral transplantation or peripheral injection of a variety of stem cells has been examined in animal models to improve the function of the aging brain.

Restoration of Telomerase Activity

Telomeres are the non-coding regions at the ends of all chromosomes in our cells. Thereby, telomeres protect the coding genes from being damaged or removed by exonuclease activity. When we are young the enzyme telomerase is active, replaces, and repairs the telomeres that are shortened by nucleases during normal cell activity. Nevertheless, as we age the production of telomerase drops down and because of the reduced activity of this enzyme the string of nucleotides in telomeres are digested without replacement. Eventually when the entire telomere is digested away, the critical coding regions are also exposed to nucleases and mutant proteins result during transcription and translation. The result is age-related diseases such as skin issues, poor eyesight, dementia, Parkinson's, Alzheimer's, cancer etc. depending on where the mutation has occurred.

Therefore, any drug or herbal that reactivates telomerase activity will result in replacing damaged telomeres and thus protect the important coding regions on the chromosomes. Recent experiments from a well-known research group in Spain showed that mice with longer telomeres showed less age-related symptoms and lived longer than other mice with normal size telomeres.

A recent paper using extracts of GOTU KOLA (*Centella asiatica*) an ancient herb used in Ayurveda reactivates telomerase activity suggesting the possibility that age-related cognitive issues such as dementia, Alzheimers, Parkinson's as well as several types of cancers can be reduced.

In this chapter, we will summarize the potential role and history of usage of herbs in treating age related conditions and potential life expectancy.

Amalaki/Amla/Indian gooseberry/Nellikkai (*Phyllanthus emblica*): Kingdom: Plantae, Clade: Angiosperms, Clade: Eudicots, Order: Malphigiales, Family: Phyllanthaceae.

This perennial tree produces segmented globular fruits that are rich in vitamin C The fruits and fruit juice are herbal treatment options for treating age-related macular degeneration and Cataract. It is a part of many polyherbal formulations.

Ashwagandha/poison gooseberry/Winter Cherry (*Withania somnifera*): Kingdom: Plantae, Clade: Angiosperms, Clade: Eudicots, Order: Solanales, Family: Solanaceae.

Although commonly used as a medicinal herb in Ayurvedic medicine, there is no conclusive clinical evidence that it is effective for treating any ailment. However, as per Ayurveda, it is prescribed for controlling stress related problems, improving cognitive brain functions, high Blood pressure, and sleeplessness. Thus, it is said to be an adaptogen.

Astragalus/huáng qí/Mongolian milk vetch (*Astragalus membranaceus/ Astragalus propinquus***: Kingdom: Plantae, Clade: Angiosperms, Clade: Eudicots, Order: Fabales, Family: Fabaceae.**

This perennial plant is a native of China. It is among the 50 most important herbals of China. Some studies show that extracts of the plant can stimulate telomerase activity and lengthen the telomeres that have shortened. Telomeres as pointed out earlier are noncoding DNA found in the ends of all chromosomes that protect the coding regions of chromosomes. Shortening of telomeres due to reduction in telomerase activity results in loss of genes. Such shortening of telomeres is a feature of the ageing cell. Thus, restoration of telomerase activity repairs and prevents the shortening of the telomeres and hence protection of the chromosomes in ageing. One of the components of Astrogalus namely **Cycloastragenol** appears to restore telomerase activity hence is useful in control of ageing.

Brahmi (*Bacopa monnieri***): Kingdom: Plantae, Clade: Angiosperms, Clade: Eudicots, Order: Lamiales, Family: Plantaginaceae**

Bacopa monnieri is a perennial, creeping herb native to the wetlands of southern and Eastern India, Australia, Europe, Africa, Asia, and North and South America. It is a memory enhancer as per Ayurvedic healers. It is considered especially useful for those who may be suffering from age-related memory loss. However, the US Food and Drug Administration (FDA) have warned manufacturers of dietary supplement products having *Bacopa monnieri* against making illegal and unproven claims that the herb can treat various diseases.

Fo-ti (*Reynoutria multiflora***) Kingdom: Plantae, Clade: Angiosperms, Clade: Eudicots, Order: Caryophyllales, Family: Polygonaceae.**

This Chinese herb is used for controlling fatigue, immune dysfunction and to treat premature aging, particularly premature grey hair. It can cause hepato-toxicity.

Ginseng (*Panax Ginseng*): Plantae, Clade: Angiosperms, Clade: Eudicots, Order: Apiales, Family: Araliaceae.

Ginseng has many phytochemicals that help stimulate and activate skin's metabolism. It is also prescribed in traditional Chinese medicine as a memory enhancing herbal.

Goji Berries/Wolf Berries (*Lycium barbarum*): Kingdom: Plantae, Clade: Angiosperms, Clade: Eudicots, Order: Solanales, Family: Solanaceae.

None of the health claims made for these berries could be verified by independent research and there have been many lawsuits against many unsubstantiated claims made by companies marketing Goji berry products filed by individuals as well as US Governmental agencies.

Indian Pennywort/Gotu Kola/Manduukaparni (*Centella asiatica*): Kingdom: Plantae, Clade: Angiospermae, Eudicots, Order: Apiales, Family: Apiaceae.

This creeping runner-herb with green rounded serrated leaves grows in the wetlands of tropical and semi tropical countries. The plants spread on the ground like a carpet and anchor themselves to the ground through rhizomes. Whole plant extracts have many phytochemicals including the triterpinoids **Asiaticoside, Brahmoside, Brahmic acid, thankunosides** as well as Pectic acid, **Hydrocotyline, vallerine**, Ascorbic acid, and various sterols. Extracts of the plant suppressed lung fibroblast proliferation in mice. Oral administration slowed the solid development of ascites tumors. Pre-treatment with plant extracts increased the survival time of irradiated animals and showed protection against radiation induced damage in liver. However, The American Cancer society does not endorse gotu kola as a treatment choice for any cancer.

Most recently, extracts of *Centella asiatica* either alone or in combination with other herbs such as Astrogalus was found to reactivate telomerase activity thereby protecting the shortening of chromosomes, which is characteristic of the ageing process.

Green Tea (*Camellia sinensis*) Kingdom: Plantae, Clade: Angiosperms, Clade: Eudicots, Order: Ericaceae, Family: Theaceae.

The young green leaves and buds that have been processed without oxidation are used to make green tea. The Tea polyphenols have a protective effect against free radicals and cardiovascular damage,

Guduchi/Giloy (*Tinospora cordifolia*): Kingdom: Plantae, Clade: Angiosperms, Clade: Eudicots, Order: Ranunculales, Family: Menispermaceae.

This shrub is native to India. The root, stems, and leaves are used for healing as per Ayurveda. The plant is commonly referred to as Amrit meaning nectar of life. The plant parts have the phytochemicals, **Columbine, tinosporaside, jatrorhizine, palmatine, berberine, tembeterine, tinocordifolioside, phenylpropene disaccharides, choline, tinosporic acid, tinosporal, and tinosporon.** The plant leaves, roots, and stem are powdered and consumed in various forms such as tonics, extracts (Kashayam), and creams. As in most Ayurvedic preparations, Guduchi is a part of many polyherbal preparations. It is often combined with Shatavari (*Asparagus*), Aloe juice or Ashwagandha.

Jiaogulan (*Gynostemma pentaphyllum*): Kingdom: Plantae, Clade: Angiosperms, Clade: Eudicots, Order: Cucurbitales, Family: Cucurbitaceae.

The fruits of this cucurbit (Squash) family are touted to have longevity properties but there is no evidence to back up these claims.

Resveratrol: skin of grapes, blueberries, raspberries, mulberries, peanuts, and other plants.

Resveratrol is a phytoalexin produced by many plants as a defense against attack by insects and other disease agents. However, it is present in substantial quantities in red wine, grapes, blueberries, raspberries, mulberries, blackberries, and peanut skins. It was initially considered as the answer to all aging problems be it cardiovascular, cancer, or diabetes. Further research has not substantiated any of these health claims and yet it is marketed by the health supplements industries vary vigorously. As of now, its value as an ant-ageing compound is questionable.

Schizandra (*Schisandra chinensis*): Kingdom: Plantae, Clade: Angiosperms, Clade: Eudicots, Order: Austrobaileyales, Family: Schisandraceae.

As per Traditional Chinese Medicine, the dried fruits of this plant are said to have many health benefits including anti-ageing properties but there is no verifiable scientific basis for these claims.

Triphala (*Emblica officinalis, Terminalia bellirica, Terminalia chebula*):

Triphala made up of the extracts of three different fruits is a major medicinal formulation in the Ayurvedic system. It is claimed to engender vibrant health and longevity by keeping balance, supporting digestion and detoxification. It is a good enhancer of bowel movements.

Turmeric (*Curcuma longa*): Kingdom: Plantae, Clade: Angiosperms, Clade: Eudicots, Order: Zingiberales, Family: Zingiberaceae.

The curcuminoids in Turmeric is claimed to have a powerful anti-ageing effect. Turmeric also has 30+ essential oils. These and other phytochemicals are claimed to have anti-inflammatory and antioxidant properties that help keep diseases at bay.

References

https://www.ncbi.nlm.nih.gov/pubmed/23432089.

https://www.sciencedirect.com/topics/agricultural-and-biological-sciences/centella-asiatica.

A. Ratz-Łyko, J. Arct, K. Pytkowska, (2016): Moisturizing and Anti-inflammatory Properties of Cosmetic Formulations Containing Centella asiatica Extract, Indian J Pharm Sci. 2016 Jan-Feb; 78(1): 27–33. PMCID: PMC4852572.

Afnan Sh. Ahmed, Muhammad Taher, Uttam Kumar Mandal, Juliana Md Jaffri, Deny Susanti, Syed Mahmood, Zainul Amiruddin Zakaria (2019): Pharmacological properties of Centella asiatica hydrogel in accelerating wound healing in rabbits, BMC Complement Altern Med. 2019; 19: 213. Published online 2019 Aug 14. doi: 10.1186/s12906-019-2625-2, PMCID: PMC6693193.

Ashok K. Shetty, Maheedhar Kodali, Raghavendra Upadhya, and Leelavathi N. Madhu, Emerging Anti-Aging Strategies - Scientific Basis and Efficacy. https://www.ncbi.nlm.nih.gov › pmc › articles › PMC6284760.

Bagherniya M, Butler AE, Barreto GE, Sahebkar A (2018): The effect of fasting or calorie restriction on autophagy induction: A review of the literature Ageing Res Rev, 47: 183–197.

Castellano JM (2018). Blood-Based Therapies to Combat Aging, Gerontology: 2019;65(1): 84-89. doi: 10.1159/000492573. Epub 2018 Sep 7.1–6.

Chung JH, Manganiello V, Dyck JR (2012). Resveratrol as a calorie restriction mimetic: therapeutic implications Trends Cell Biol, 22: 546–554.

Cox KH, Pipingas A, Scholey AB (2015). Investigation of the effects of solid lipid curcumin on cognition and mood in a healthy older population J Psychopharmacol, 29: 642–651.

Gao Y, Wei Y, Wang Y, Gao F, Chen Z (2017). Lycium Barbarum: A Traditional Chinese Herb and A Promising Anti-Aging Agent Aging Dis, 8: 778–791.

Gurau F, Baldoni S, Prattichizzo F, Espinosa E, Amenta F, Procopio AD, et al. (2018). Anti-senescence compounds: A potential nutraceutical approach to healthy aging Ageing Res Rev, 46: 14–31.

Hanjani NA, Vafa M (2018). Protein Restriction, Epigenetic Diet, Intermittent Fasting as New Approaches for Preventing Age-associated Diseases Int J Prev Med, 9: 58.

Hansen M, Rubinsztein DC, Walker DW (2018) Autophagy as a promoter of longevity: insights from model organisms Nat Rev Mol Cell Biol, 19: 579–593.

Harman D (1992). Free radical theory of aging: history, EXS, 62: 1–10.

Kamonwan Jenwitheesuk, Porntip Rojsanga, Bowornsilp Chowchuen, Palakorn Surakunprapha A (2018): Prospective Randomized, Controlled, Double-Blind Trial of the Efficacy Using Centella Cream for Scar Improvement, Evid Based Complement Alternat Med. 2018: 9525624. Published online 2018 Sep 17. doi: 10.1155/2018/9525624, PMCID: PMC6166374.

Kenyon CJ (2010). The genetics of ageing Nature, 464: 504–512.

Kirk-Sanchez NJ, McGough EL (2014). Physical exercise and cognitive performance in the elderly: current perspectives. Clin Interv Aging, 9: 51–62.

Kloska D, Kopacz A, Piechota-Polanczyk A, Nowak WN, Liguori I, Russo G, Curcio F, Bulli G, Aran L, Della-Morte D, et al. (2018). Oxidative stress, aging, and diseases Clin Interv Aging, 13: 757–772.

Li L, Wang Z, Zuo Z (2013). Chronic intermittent fasting improves cognitive functions and brain structures in mice. PLoS One, 8: e66069.

Liu P, Zhao H, Luo Y (2017). Anti-Aging Implications of Astragalus Membranaceus (Huangqi): A Well-Known Chinese Tonic. Aging Dis, 8: 868–886.

Mattson MP, Longo VD, Harvie M (2017). Impact of intermittent fasting on health and disease processes Ageing Res Rev, 39: 46–58.

Mosher KI, Luo J, et al. (2014): Young blood reverses age-related impairments in cognitive function and synaptic plasticity in mice. Nat Med, 20: 659–663.

Most J, Tosti V, Redman LM, Fontana L (2017): Calorie restriction in humans: An update. Ageing Res Rev, 39: 36–45.

Park D, Yang G, Bae DK, Lee SH, Yang YH, Kyung J, et al. (2013): Human adipose tissue-derived mesenchymal stem cells improve cognitive function and physical activity in ageing mice. J Neurosci Res, 91: 660–670.

Radak Z, Torma F, Berkes I, Goto S, Mimura T, Posa A, et al. (2018). Exercise effects on physiological function during aging. Free Radic Biol Med, S0891-5849(18): 32273–1.

Sadowska-Bartosz I, Bartosz G (2014): Effect of antioxidants supplementation on aging and longevity. Biomed Res Int, 2014: 404680.

Schmitt R (2017). Senotherapy: growing old and staying young? Pflugers Arch, 469: 1051–1059.

Scudellari M (2015): Ageing research: Blood to blood, Nature, 517: 426–429.

Small GW, Siddarth P, Li Z, Miller KJ, Ercoli L, Emerson ND, et al. (2018). Memory and Brain Amyloid and Tau Effects of a Bioavailable Form of Curcumin in Non-Demented Adults: A Double Blind, Placebo-Controlled 18-Month Trial. Am. J. Geriatr Psychiatry, 26: 266–277.

Sun N, Youle RJ, Finkel T (2016): The Mitochondrial Basis of Aging. Mol Cell, 61: 654–666.

Theodotou M, Fokianos K, Mouzouridou A, Konstantinou C, Aristotelous A, Prodromou D, et al. (2017). The effect of resveratrol on hypertension: A clinical trial. Exp Ther Med, 13: 295–301.

Tiwari SK, Agarwal S, Seth B, Yadav A, Nair S, Bhatnagar P, et al. (2014). Curcumin-loaded nanoparticles potently induce adult neurogenesis and reverse cognitive deficits in Alzheimer's disease model via canonical Wnt/beta-catenin pathway. ACS Nano, 8: 76–103.

van Praag H, Christie BR, Sejnowski TJ, Gage FH (1999): Running enhances neurogenesis, learning, and long-term potentiation in mice. Proc Natl Acad Sci U S A, 96: 13427–13431.

Wiesława Bylka, Paulina Znajdek-Awiżeń, Elżbieta Studzińska-Sroka, Małgorzata Brzezińska (2013): Centella asiatica in cosmetology, Dermatol Alergol. 2013 Feb; 30(1): 46–49. Published online 2013 Feb 20. doi: 10.5114/pdia.2013.33378, PMCID: PMC3834700.

Wei W, Ji S (2018). Cellular senescence: Molecular mechanisms and pathogenicity. J Cell Physiol, 233: 9121–9135.

Yang Y, Ren C, Zhang Y, Wu X (2017): Ginseng: A Nonnegligible Natural Remedy for Healthy Aging. Aging Dis, 8: 708–720.

Yousefzadeh MJ, Zhu Y, McGowan SJ, Angelini L, Fuhrmann-Stroissnigg H, Xu M, et al. (2018). Fisetin is a senotherapeutic that extends health and lifespan. EBioMedicine, 36: 18–28.

Zhang Y, Kim MS, Jia B, Yan J, Zuniga-Hertz JP, Han C, et al. (2017). Hypothalamic stem cells control ageing speed partly through exosomal miRNAs. Nature, 548: 52–57.

Zhao H, Luo Y (2017). Traditional Chinese Medicine and Aging Intervention, Aging Dis, 8: 688–690.

Zhao HP, Han ZP, Li GW, Zhang SJ, Luo YM (2017). Therapeutic potential and cellular mechanisms of panax notoginseng on prevention of aging and cell senescence-associated diseases, Aging Dis, 8: 721-739.

Zuo W, Yan F, Zhang B, Li J, Mei D (2017). Advances in the Studies of Ginkgo Biloba Leaves Extract on Aging-Related Diseases, Aging Dis, 8: 812–826.

CHAPTER 14

Alphabetical List of Medicinal Herbs and Phytochemicals

Introduction

The Biology of various important conditions that affect human health as well as potentially important plants that help to treat these conditions were described in the earlier chapters. In this Chapter, an alphabetical list of the most important plants and phytochemicals that have real or potential medicinal properties and are prescribed by practitioners of traditional medicine is listed and briefly discussed. A fuller description of many of these plants is given in the proper chapter and the reader is requested to refer to these chapters for fuller descriptions. Although individual herb extracts in the form of powders, pastes, capsules, tablets or tonics and drinks are prescribed for certain conditions, it is more common to formulate polyherbal formulations having several herbal components. It must be noted that all herbals and herbal formulations are often based on anecdotal and historical basis and only a few have undergone modern clinical tests. But many people in the world still do not have access to modern clinically proven medicines and are dependent on herbalists for treatment. On the other hand, many who do have access to modern medicine are turning increasingly to herbal medicines and supplements for both preventative as well as treatment options for supporting good health.

The plants and phytochemicals listed below have multiple properties but have been chiefly associated with the medical conditions described in earlier chapters. A few of the plants listed earlier are not listed here. While individual phytochemicals have specific properties, herbalists consider that it is a combination of different phytochemicals through individual and synergistic effects in a plant extract that has the desired medicinal effect. It must be pointed out that not all these herbals have gone through modern protocols of testing in clinical settings. Usually only individual components of a potential medicinal herb are assessed by modern statistically relevant protocols but not the totality of phytochemicals found in the plant.

Abarema (*Abarema sp.*): There are 124 species of Abarema listed in the Plant list. They are natives of Amazonian forests, particularly Brazil. *Abarema cochliacarpos* extracts are used to counteract snake venom. Extracts of the same species are also reputed to heal gastric ulcer and gastritis.

Abuta/Midwifes herb (*Cissampelos pareira*): Abuta is a Medicinal herb that helps to purify blood and support hormone levels in the body. It is commonly known as 'midwife's herb' as it can be used to treat a wide variety of women's complaints, such as menstrual cramping and minor reproductive tract conditions.

Acalypha (*Acalypha integrifolia*): Leaf extracts and infusions have anti parasite properties and hence used as an herbal for treating intestinal worms.

Acacetin – (Damiana-Turnera diffusa, Black locust (*Robinia pseudoacacia*), Silver birch (*Betula pendula*). This flavanoid is an aromatase inhibitor. It reduces estrogen and increases testosterone. It is an anti-inflammatory agent and is also considered to protect the heart. It has no side effects

African Cherry (*Pygeum Africanum/Prunus Africanum*): Extract from the African tree bark is one of the components of many prostate health product supplements. But there are many reports that discard its usefulness in treating benign prostate hyperplasia or enlarged prostate (BPH).

African cherry orange/Omuboro (*Citropsis articulata*): In Uganda, an infusion made of the ground root of the Omuboro, drunk once a day for three days is considered powerful aphrodisiac for men only.

African Custard Apple/Wild Custard Apple/African Custard Apple/ Wild Custard Apple/Soursop (*Annona senegalensis*): Bark and seed powders have insecticidal properties and are used in herbal formulations for treating parasitic intestinal infestations.

African dream root (*Silene capensis*): Root extracts are sedatives and induce vivid dreams after smoking.

Ajwain/Ajwain/Carum/Bishops weed/Omum, (*Trachyspermum ammi/ Carum copticum/Ammi copticum*): Chewing raw Carum seeds (alone or with a little sugar) can help with digestion. Ajwain also has properties that help reduce the flatulence-causing effect of beans, and has been shown to cure diarrhea, dysentery, and indigestion when made into a "tea.

Alligator Yam, Milky yam/Vidari kanda (*Ipomoea digitata/I. paniculata/I. mauritiana*): The underground tuber extracts is mixed in milk for treating menstrual irregularities. It is also used to boost urination, stimulate the digestive system, induce menstruation and lactation, bile movement, boost sexual desire, and increase metabolic rate.

Aloe (*Aloe vera*): A. vera contains phytochemicals such as pyrocatechol, saponin, Acemannan, β-mannan, anthraquinone, C-glycosides, diethylhexylphthalate, bradykininase, oleic acid, phytol, and magnesium lactate along with water-soluble polysaccharides like glucomannan. Reports indicate that Aloe has been used successfully to treat burns and skin irritations, canker sores (aphthous stomatitis), gastric and duodenal ulcers caused by *Helicobacter pylori*. *Aloe vera* polysaccharides hold water and help in keeping the skin moist and hydrated. It is a part in many moisturizing gel preparations for skin. It helps in healing skin abrasions, eczema irritation and sun burn. The dried sap of the aloe vera plant has traditionally been used in Arabia to treat diabetes. Other studies show that the juice from the *Aloe vera* plant can help lower blood sugar in people with types 2 diabetes. One of the ingredients **Acemannan** is an immune booster. Other ingredients include Anthro-quinones (Alloin A and B) which manage the laxative properties of Aloe juice.

Ashoka Tree (*Sarco asoca/Saraca indica*): Extracts of the bark are used as topical applications to improve skin health. It is used to control psoriasis, eczema, skin irritations and other skin conditions. Bark powders also control gastric irritation.

Amalaki/Amla/Indian gooseberry/Nellikkai (*Phyllanthus emblica/ Embelica officianalis*): The extracts from various parts of *E. officinalis*, especially fruit, contain numerous phytoconstituents such as higher amount of polyphenols like gallic acid, ellagic acid, different tannins, minerals, vitamin C, amino acids, fixed oils, and flavonoids like rutin and quercetin. The extract of plant is reported to be efficacious against Diabetes, inflammation, cancer, osteoporosis, neurological disorders, hypertension together with lifestyle diseases, parasitic and other infectious disorders. The fruits from this perennial tree prevents or reduces macular degeneration and Cataract. It is a part of many polyherbal formulations including Triphala.

Ammi/Khella/toothpick weed (*Ammi visnaga*): Extracts of this plant is used in the middle east as a treatment for expelling kidney stones. It also contains Khellin which controls aberrant heart rhythms and stimulates melanocytes in vitiligo patients. It has many side effects including potential for skin cancers and hear attacks.

Andrographis/Green Chiretta (*Andrographis paniculata*): The leaves and roots of this member of the Lamiaceae family is used to treat bacterial infections as well as jaundice. It is reported to have anti-viral properties and is a popular ingredient of Ayurvedic and Siddha forms of herbal medications.

Angelica (*Angelica archangelica*): The root, seed, leaf, and fruit are used to make medicine. Angelica is used for treating heartburn intestinal gas, loss of appetite, overnight urination, arthritis, stroke, dementia, circulation problems, "runny nose", nervousness and anxiety, fever, plague, and trouble sleeping. It is also used for treating Nicotine withdrawal.

Anise: (*Pimpinella anisum*): Anise seeds are used for treating upset stomach, intestinal gas, runny nose, and as an expectorant to increase productive cough. It is also used to increase urination and to stimulate the appetite. Women use anise to increase milk flow when nursing, start menstruation, treat menstrual discomfort or pain, ease childbirth, and increase sex drive. Men use anise to treat symptoms of "male menopause." Other uses include treatment of seizures, nicotine dependence, trouble sleeping, asthma, diabetes, and constipation.

A close relative known as star Anise is a rich source of shikimic acid which is used for producing the anti-flu virus drug Tamiflu. But currently Tamiflu is produced through recombinant fermentation of E. Coli technology.

Anthocyanins and Anthocyanidins (Found in flowers and colored parts of plants): There are no substantial clinical trials showing that dietary anthocyanins have any beneficial physiological effect in humans or lower the risk of any human diseases.

Anthraquinones: These phytochemicals are powerful laxatives found in *Aloe vera*, Rhubarb, *Cascara sagrada* and a few other plants. They also have anti-bacterial properties. Long time usage of plants having Anthraquinones cause the reversible darkening of the lining of the colon. Extracts of plants having this chemical are marketed as laxatives in health food stores but is not an FDA approved drug.

Apigenin – (Parsley-*Petroselinum crispum*, Celery-*Apium graveolens*, Celeriac- *Apium graveolens* var. rapaceum, Chamomile -*Matricaria recutita,*/chamomile -*Chamaemelum nobile*, Brahmi-*Bacopa monnieri*). The bioflavonoid Apigenin reduces anxiety and causes sedation.

Apple (*Pyrus malus*) Cider Vinegar: Some studies show that consumption of Apple Cider vinegar decreases fasting sugar and helps to lower A1c levels slightly.

Arjuna: (*Terminalia Arjuna*): The bark extracts have a cardioprotective role by restoring the depleted endogenous myocardial antioxidants and improving myocardial function. The bark contains: Phenolic compounds – Terminic acid and Arjunolic acid, Glycosides: Arjunetin and Arjunosides I–IV, Flavones, Tannins, Oligomeric Proanthocyanidins, B-Sitosterol, Casuarinin, Phenolic acids like ellagic acid and gallic acid and Lactones. The plant extracts also control polyurea, heals ulcers and bleeding.

Artemesia/Wormwood/Sweet Annie (*Artemesia annua*): Extracts and oil from all above parts of plant are used as herbal medicine. The plant has Artemisinin a powerful antimalarial agent. Besides, plant extracts in the form of tea are good for control of intestinal spasms and control of diarrhea. In Mexico, Artemesia leaf infusions are a common remedy for controlling diabetes.

Artichoke flowers *(Cyanara scolymus)*: Artichoke or Globe artichoke (*Cynara cardunculus var. scolymus*): Artichoke is a powerful antioxidant. Artichoke has been traditionally used as diuretic, liver tonic and other medical conditions such as: stomach acidity, irritable bowel syndrome symptoms, and even eczema (flower paste).

Asafoetida/Hing/Perumkaayam (*Ferula asafoetida*): The gum-resin of this plant is used for treating stomach pain and in Ayurvedic medicine, it is used to aid digestion and gas, as well as treat bronchitis and kidney stones. There are also reports that it increases sperm count and has aphrodisiac properties.

Ashwagandha/Indian Ginseng/Poison gooseberry/winter cherry (*Withania somnifera*): The phytochemicals in Withania handle control of tremors associated with Parkinson's disease. It is recommended by herbalists for improving cognition. Withania confers relief from stress and helps to sedate and allow good sleep.

Asparagus/Shatavari (*Asparagus racemosa*): Women use the dried root extracts for controlling premenstrual syndrome (PMS) and uterine bleeding and to start breast milk production. *Asparagus racemosa* is also used to increase sexual desire as an aphrodisiac. This plant is different from Asparagus vegetable which is *Asparagus officianalis*. Asparagus vegetable is also considered to be an aphrodisiac and diuretic. However, the urine passed by people who eat Asparagus shoot vegetable has a strong foul smell.

Asthma herb/Amman Pacharisi/Malnommee (*Euphorbia hirta*): A common weed plant in India and Africa which produces latex upon wounding, the latex when applied to warts softens the warts which eventually fall off in a few days. The juice from the plants is a medicine for Asthma, bronchitis, antidiabetic and anti-amoebic. Extracts are marketed as antiamibien in Europe and is listed in the African pharmacopoeia. It also has antibacterial and anti- inflammatory properties.

Astragalus/huáng qí/Mongolian milk vetch (*Astragalus membranaceus/ Astragalus propinquus*): Extracts of this plant named TA-65 is believed to prevent shortening of the ends of chromosomes and thereby prevent loss of important genes. Many scientific reports show that shortening of chromosome ends known as telomeres is a regular feature of ageing. Thus, any medication that prevents shortening of chromosomes will reduce the effects of ageing and age-related diseases.

Atropine – Belladonna (*Atropa belladonna*), Jimson weed (*Datura stramonium*), mandrake (*Mandragora officinarum*): Atropine is a medication used to treat certain types of poisonings as well as some types of slow heart rate, and to decrease saliva production during surgery. It is given intravenously or intramuscularly. Eye drops are used to treat uveitis and early amblyopia. Belladonna plasters are used for reducing back spasms, Side effects are dry mouth, large pupils, urinary retention, constipation, and a fast heart rate.

Babchi (*Psoralea corylifolia*): Extracts of this plant are used in skin formulations for treatment of vitiligo. The plant extracts in combination with photoactivation slows down the degradation of the pigment producing melanocytes and thus reduce the vitiligo symptoms.

Balloon vine/Valli Uzhinja (*Cardiospermum halicacabum*): Extracts of the leaves and vine controls skin irritation and inflammations. When used as a hair shampoo, it is said to control dandruff and premature greying of hair. The leaf juice is prescribed for control of earaches.

Barberry/Berberine (*Berberis vulgaris*): Studies suggest that supplemental berberine and barberry extract can lower your blood sugar levels and may help treat diabetes.

Basil/Holy Basil (*Ocimum tenuiflorum/O. sanctum*): The holy Basil known as Thulasi in India are of two varieties namely green and red. Both have several medicinal properties. The leaf extracts and the aromatic oil from leaves have antimicrobial, antidiarrheal, antioxidant, anti-inflammatory, anti-diabetic, anti-pyretic (reducing fever), antitussive (treats cough), anti-arthritic (controls arthritis), anti-coagulant, apoptogenic (promoting programmed cell death-an important feature of cancers) properties.

Beach (*Moonflower Ipomoea*): The grounded seeds are used to make an infusion, which is used to induce a numbing trance for the purpose of divination and diagnosis.

Berberine – California poppy, Oregon grape, barberry, and golden seal: Berberine is prescribed in Chinese medicine for control of bacterial infections of the GI tract. Oral Berberine is recommended for treating diabetes and cholesterol and berberine topicals are prescribed for treating cankers and sores. It is also used in eye drops to control trachoma. Taking berberine products along with insulin might cause hypoglycemia.

Betulinic acid (White birch-*Betula pubescens*, Indian jujube -*Ziziphus mauritiana*, self-heal -*Prunella vulgaris*, rosemary -*Rosmarinus officinalis*, Chinese anemone root -*Pulsatilla chinensis*): Betulinic acid has potential as an anticancer agent, by inhibition of topoisomerase.

Betanin (Red beet root-*Beta vulgaris*): This is a Betalain pigment which has anti-microbial properties. Beet root powder is being marketed as a powerful antioxidant and as a cardio protective agent but without any real basis.

Bilberries/European Blue berries/Whortleberry (*Vaccinium* sp.): This plant is related to blue berries and cranberries. Bilberries are rich in anthocyanins like Cyanidin, Delphinidin, Malvidin, Peonidin and Petunidin. Berries and berry extracts are used for supporting good eye health and prevent onset of glaucoma, cataract, eye strain and stress.

Traditionally, bilberry leaves have been used to control blood sugar levels in people with diabetes. Research shows that all berries help reduce the body's glucose response after eating a high sugar meal. European health care professionals use bilberry extracts to control venous insufficiency, which occurs when valves in veins in the legs that carry blood to the heart are damaged. Studies have reported improvements in symptoms, but most studies were poorly designed. Bilberry has multiple uses as a dietary supplement for cardiovascular conditions, diarrhea, urinary tract infections, eye problems, diabetes, and other conditions. Bilberry extract is sold in tablets, capsules, and drops, and the berries are sold dried and as a powder. The leaves are made into teas. Eating the fruits is considered to confer good circulatory effects, reduce oxidative stress, and generate anti-inflammatory activities. Some studies also show that it helps to control diabetes and hypertension-induced retinopathy.

Bitter gourd/Bitter melon (*Momordica charantia*): Bitter melon is a popular ingredient of Asian cooking and Indian and traditional Chinese medicine. It is believed to relieve thirst and fatigue, which are symptoms of type 2 diabetes. Research has shown that extract of bitter melon can reduce blood sugar. But in addition, it is also an abortifacient and hence should be avoided by pregnant women. Excess consumption of bitter gourd for long periods could lead to adverse health effects. These include miscarriage, drug interactions, liver inflammation, and irregular heart rhythm.

Black bean (*Castanospermum austral*): Crude extract of C. austral has been reported to cause a fall in systolic as well as diastolic blood pressure in a dose-dependent manner.

Black cohosh/black bugbane/Black snakeroot/fairy candle (*Actaea racemosa*): The dried root extracts are used for treating female health issues relating to control of hot flashes and other menstrual issues.

Black muesli/Kali musli/Nilappana: (*Curculigo orchioides*): This plant fruits are regarded as a potent adaptogen and aphrodisiac in Ayurvedic system of medicine. It is an important ingredient of many Ayurvedic preparations and is considered to have aphrodisiac, immunostimulant, hepatoprotective, antioxidant, anticancer and antidiabetic activities.

Black plum/Malabar Plum/Jambolan/(*Syzygium cumini* also named as *Eugenia jambolana* or *E. Cumini*): The bark, Leaves, fruits, and seed extracts are commonly used in herbal medicines in Indian sub-continent as well as in Australasia. The plant has been reported to have ellagic acid, triterpenoids, acetyl oleanolic acid, quercetin, isoquercitin, myricetin and kaempferol. The fruits have vitamin C as well as small quantities of other vitamins and minerals. *S. cumini* possesses anti diabetic, antimicrobial, hypolipidemic (reducing cholesterol), anti-allergic, anti-inflammatory, cardio-protective (heart), hepato-protective(liver) and anti-cancer properties. Thus, it is plant with multipurpose properties and is an important medicinal plant.

Blond Psyllium/Isabgol/Indian Plantago (*Plantago ovata*): Psyllium plant fiber from seed husk is found in common bulk laxatives and fiber supplements. Psyllium has also been used historically to treat diabetes.

Studies show that people with type 2 diabetes who take 10 grams of psyllium every day can improve their blood sugar and lower blood cholesterol.

Blue berries (*Vaccinium sp.*): Fruits are extremely rich in flavones, anthocyanins, and antioxidants, which make the fruits ideal for conferring many health benefits including cardio-protection.

Boldine – Boldo, Evergreen lindera: Boldine is an alkaloid of the aporphine class. It has antioxidant properties and is reported to have anti-inflammatory, liver protection and fever reduction properties.

Brahmi/Indian Pennywort (*Bacopa monnieri*): Pennywort or Bacopa extracts are prescribed for improving cognitive functions and memory retention. Powders of bark and leaves enhance memory and thus help prevent memory loss.

Bringaraj/Karisilankanni (*Eclipta alba*): This plant belonging to the Sunflower family grows in temperate to tropical areas. Extracts of the plant are sed to make herbal hair -oils and creams. A handful of chopped leaves is added to vegetable dishes during cooking or to other food items as a general medicinal tonic.

Buchu (*Agathosma betulina*-Family: Rutaceae): It is a South African medicinal plant and has been used by the indigenous people of the area for centuries to treat many ailments. It is an effective diuretic and anti-inflammatory agent. Early Dutch settlers used Buchu to make a brandy tincture, which is still used today to treat many disorders.

Burdock (*Arctium lappa/A. sp.*): Burdock root oil extract is used for scalp treatment. The tap roots are used in culinary preparations in China, Korea, and Japan.

Bush/Ceylon leadwort/Wild leadwort/Chitraka-Chitramol in Sanskrit, Koduveli in Tamil-India (*Plumbago zeylanica* L): This plant is a native of Africa and Indian subcontinent. It is a multipurpose medicinal plant. Plant extracts are effective against bacterial, fungal, intestinal worms like hook worm infections. Additionally, it controls skin infections due to scabies, ringworm, dermatitis, acne, sores, and ulcers. Extracts of the plant are important constituents of various Ayurvedic, Siddha and Unani herbal formulations.

Caffeine – Coffee bean from *Coffea arabica*: Caffeine stimulates the nervous system, reduces appetite, and keeps people awake.

Calendula/Pot Marigold/Scotch marigold (*Calendula officianalis***):** The natural oil extracted from Calendula flowers is used to control diaper rash, eczema, athletes foot fungus and other skin infections.

California poppy (*Eschscholzia californica***):** The whole plant, especially the flowers and the leaves, is anti-inflammatory, antiseptic, and antispasmodic. It helps discharging bile, increases perspiration, emmenagogue (stimulating blood flow to pelvic region), and helps the healing process of wounds. Leaves and root extracts are reputed to have sedative effects and as such teas made of these extracts are consumed to induce sleep.

Camellia (*Camellia sinensis***):** Green tea was shown to have analgesic and anti-inflammatory properties and may constitute a natural treatment choice in chronic inflammatory disorders.

Camphor tree (*Cinnamomum camphora***):** Camphor from this plant is a part of many analgesic preparations. Some studies have shown that camphor oil can repair some of the damage caused by excessive exposure to UV light. It has been found also to help in healing of secondary burn injuries.

Cannabis/Marijuana//Charas/Ganja/pot (*Cannabis sativa* **ssp. sativa;** *C. sativa* **ssp. indica,** *C. Sativa***:** All three sub-species and varieties have cannabinoids. Tetra hydro cannabinol (THC) and Cannabidiol (CBD) have analgesic properties besides several other medicinal properties. Cannabis leaf/flower extracts/seed oil are all reported to improve good skin health.

Cannabis web sites and pharmacies recommend the following strains for control of Seizures: 1. Indica varieties: Argyle, Athabasca, Bell ringer, Big Bang, Black Bubba, Brandywine, Darkside, Devils fruit, Fin Lou Zer, Gorilla Biscuit, Grape Ape 2. Sativa varieties: Lemon Pie and Southern Lights,3. Hybrids: Aloha berry, Blackberry dream, Bedford Glue, Dreamers Glass,

Frisian dew, Frosted Freak, Green Crack, GI001, G.O./A.T., Marcosus Marshmallow, Maui Haole, Red Dragon, White Widow.

The cannabinoids in this plant especially CBD is reported to be beneficial in calming patients and controlling tremors.

The Cannabis variety charlotte's web which is rich in CBD but has only traces of Tetrahydro cannabidiol (THC) was experimentally found to control a type of epilepsy known as Dravet syndrome. Since then, the US Food and Drug Administration (FDA) approved Epidiolex (cannabidiol as the therapeutic ingredient) oral solution for the treatment of seizures in two forms of epilepsy, Lennox-Gastaut syndrome and Dravet syndrome, for children two years of age and older.

Capsicum annum (Hot/Chili pepper): Extracts of the fruits have capsaicin which has pain control properties. Capsaicin is a part of many muscle rub preparations.

Capsaicin – Chili pepper (*Capsicum* spp.). The phytochemical Capsaicin is used to help pain of neuralgia (pain in the nerves) due to different reasons including shingles and minor pain associated with rheumatoid arthritis or muscle sprains and strains.

Caraway/Kala Jeera/Perum Jeerakam (*Carum carvi*): The oil, fruit, and seeds are used as medicine for digestive problems including heartburn, bloating, gas, loss of appetite, and mild spasms of the stomach and intestines. Caraway oil is also taken by mouth to help people cough up phlegm, improve control of urination, kill bacteria in the body, and relieve constipation. Caraway oil is used in mouth washes.

Cardamom (*Eletteria cardamom*): Cardamom seeds and seed oil are used as flavoring agents and for various medical conditions. It is a good mouth freshener, relieves bloating and promotes digestion.

Carotenoids -Tetraterpenoids: Carotenoids such as lutein and zeaxanthin, are known as xanthophylls. The oxygen free carotenoids such as α-carotene, β-carotene, and lycopene, are known as carotenes. Lutein is found in Nasturtium, dandelion leaves, spinach, and kale and in yellow carrots and turnips. It is a popular supplement purported to help vision.

Neither the U.S. Food and Drug Administration nor the European Food Safety Authority considers lutein an essential nutrient or has acted to set a tolerable upper intake level. Zeaxanthin found in the leaves of every green plant is also recommended as a supplement to help improve eye vision. The actual health effects of zeaxanthin and lutein are not proven.

Castor beans (*Ricinus communis*): Castor oil expressed from seeds is a well-known purgative and is commonly used to induce bowel movements, expelling worms, and cleansing of intestine. The seeds hold an extremely powerful cytotoxic agent namely ricin.

Cat's Claw (*Uncaria rhynchophylla*): Cats claw extracts have been used to lower BP and to relieve various neurological symptoms. The hypotensive activity has been attributed to an indole alkaloid called hirsutine.

Catechin hydrate (*Acacia catechu* extract): Catechin is a polyphenolic flavonoid from tea leaves, grape seeds, and the wood and bark of trees such as acacia and mahogany. Catechin is reported to suppress the effects of Parkinson's and Alzheimer's disease.

Catharanthus (*Catharanthus trichophyllus*): This plant has lower concentrations of Vinca alkaloids. It is prescribed as an aphrodisiac and to treat male impotence. It is also an appetite suppressant. Look also under Vinca or Madagascar Periwinkle.

Cathine – Khat/*Catha edulis*: -Norpseudoephedrine, also known as cathine, is a psychoactive drug which acts as a stimulant.

Catnip (*Nepeta cataria*): Leaves have the terpenoid nepetalactone which induces sedation but also hyperactivity in cats. It is often a suggested ingredient of sleep preparations.

Catuaba/Caramuru (*Erythroxylum catuaba/vaccinifolium*): People use catuaba for sexual arousal and performance, agitation, poor memory, and many other conditions.

Celery (*Apium graveolens*-Apiaceae): As per traditional Chinese Medicine, Celery is effective for hypertension because it acts upon the liver. Fresh celery juice mixed with vinegar is used to relieve dizziness and headache and shoulder pain associated with high Blood pressure.

Chaff-flower/prickly chaff flower/devil's horsewhip/Nayuruvi/ Apaamaarga- (*Achyranthus aspera* Lin): Root extracts are used as aphrodisiac and sex tonics and as sedative, given for inflammation of mouth membranes(stomatitis), fatty stools (steatorrhea), sterility and impotence. The root powder is taken orally to check discharge of sperm with urine and to increase sexual potentiality.

Chamomile (*Matricaria recutita*): Chamomile tea infusions induce sleep. The tea infusions also have antioxidant properties.

Chanca Piedra/Kidney stone tree'/Kizha Nelli (*Phyllanthus niruri, P. quebra*): The effect of chanca piedra in people with kidney stones is unclear. Research results have been conflicting. But it is alleged that taking daily doses of the whole plant extract of this species reduces the size of stones and allows easy discharge of kidney stones through urine. The plant extracts are also prescribed for control of liver inflammation.

Chickweed (*Stellaria media*): Chickweed cream or lotion helps in soothing the irritation caused by skin disorders such as psoriasis and eczema.

Chinese Salvia/Danshen/Chinese red Sage (*Salvia miltiorrhiza*): The plant parts are used to treat Chest pain (angina), high cholesterol, high blood pressure and ischemic stroke.

Chinese Hawthorn (*Crataegus pinnatifida*): It has been used in China as a decoction for treatment of hypertension for thousands of years. Pharmacological and clinical trials have shown that it lowers BP. The two main substances that contribute to hawthorn's beneficial effects on heart are flavonoids and oligomeric procyanidins, which are potent antioxidant agents.

Chirayita/Kirata/Indian Gentian (*Swertia chirayita*): Leaf extracts and powders are used in decoctions, tonics to stimulate good liver function, optimal flow of bile, improve digestion, control hiccups, and treat viral and bacterial diseases. It is a part of the Ayurvedic medicine Sudarshana Churna prescribed for treating liver diseases, stomach function and general tonic.

Chrysin – (Purple passionflower-*Passiflora incarnata*, Blue passionflower-*Passiflora caerulea*, chamomile -*Matricaria recutita, Chamaemelum* *nobile,* **midnight horror-*Oroxylum indicum*). Chrysin is a flavonoid that is used in body building formulations, erectile dysfunction, and stress relief.

Cinnamon-Ceylon Cinnamon (*Cinnamomum verum/C. Zeylanicum***):** The herb is recommended for improving peripheral blood flow. However, recent studies show that it has insulin like properties by stabilizing blood sugar levels in people with mild type 2 diabetes. Consuming about half a teaspoon of cinnamon per day can result in significant improvement in blood sugar, cholesterol, and triglyceride levels in people with type 2 diabetes.

Citrus rind and fruits (*Citrus* **sp.):** Citrus rind and juice have Hesperidin and Diosmin which are two bioflavonoids that are found to improve vein health. When consumed in the form of capsules or smeared on the legs they improve venous health by reducing swelling and improving circulation. These two flavanoids are used for treating varicose veins, spider veins and hemorrhoids. The essential oil and fruit extracts have furano coumarins which are photosensitive and can cause photo dermatitis when skin smeared with orange extracts are exposed to light for prolonged times.

Chlorogenic acid (Roselle-*Hibiscus subdariffa,* **Eggplants, peaches, prunes, coffee beans):** Oral ingestion reduces blood pressure and has anti inflammation properties.

Cloves (*Syzygium aromaticum***):** Clove (*Syzygium aromaticum* flower buds) ethanolic extracts significantly suppressed an increase in blood glucose level in type 2 diabetic mice. These and other results show that clove has potential as a functional food ingredient for the prevention of type 2 diabetes.

Coca (*Erythroxylum coca***):** The leaves of this plant have the potent drug cocaine. The leaves and bark of this plant are chewed by the people living in the Andes region of South America for over 3000 years as a herbal medicine to control hunger, increase stamina, control stomach pains, depression, high altitude symptoms and body pain. A derivative-lidocaine is used as a local anesthetic. Modern medicine has focused on cocaine although it has other alkaloids such as benzoylecgonine, ecgonine, tropacocaine, hygrine, cuscohygrine, and nicotine. Cocaine is a very addictive drug and is a powerful central nervous system simulator.

Cocaine – Coca Leaves/*Erythroxylum coca* and *Erythroxylum novogranatense*: Cocaine controls pain and can stimulate the body and reduce hunger. It is addictive.

Cocoa/chocolate (*Theobroma cacao*): Seeds are rich in flavanoids and antioxidants. Cocoa and chocolate products are helpful in protecting the cardiovascular system. Cocoa and chocolate help reduce inflammation, reduce risks of stroke, and improve cardio circulation.

Coconut (*Cocos nucifera*): Coconut water induces urine flow and, in the process, helps to evacuate kidney stones. Coconut water has many electrolytes and as such control's symptoms of electrolyte loss due to perspiration. During World War II, the coconut water was used as an intravenous supplement in the absence of glucose-saline.

Coffee weed (*Cassia occidentalis*): This is a small tree growing 5 to 8 m in height. The leaf of this plant is used in local folk medicine as an anti-high blood pressure agent.

Colchicine/Autumn crocus/(*Colchicum autumnale*): Colchicine is a mutagen and is used to control gout. This phytochemical is approved by FDA for control of gout.

Comfrey/knit bone/bonese (*Symphytum uplandica* or *S. officinale*): Comfrey extracts applied topically is prescribed for healing broken bones and joints. The FDA has banned products having Comfrey for internal use.

Common cabbage Tree (*Cussonia spicata*): Decoctions from bark are used for treating malaria. Root extracts are prescribed for treating fever, diarrhea, and venereal disease.

Common Sage (*Salvia officinalis*): In some studies, supplementation of extracts of the common sage leaves which holds antioxidants led to a reduction in cognitive problems and an improvement in attention.

Copaibo/Palo de aceite (*Copaifera officinalis*): Copaibo balsam and oil are prescribed for treating urinary tract infections but the data is not convincing.

Coumarin – (Tonka bean -*Dipteryx odorata*, vanilla grass -*Anthoxanthum odoratum*, sweet woodruff-*Galium odoratum*, great mullein-*Verbascum thapsus*, sweet grass-*Hierochloe odorata*, Cassia cinnamon-*Cinnamomum cassia*, deertongue -*Dichanthelium clandestinum*, Criollo tea -*Justicia pectoralis*). Coumarins are anti coagulants and as such foods having coumarins can antagonize vitamin K, hence clotting. Such foods should not be used in patients already on blood thinning agents.

Cow-Itch Plant/Cowhage/Velvet bean (*Mucuna pruriens/Mucuna prurita*): Seed powders or other extracts have the chemical known as levodopa which is the allopathic chemical used in treating Parkinson's disease. Apart from Parkinson's, extracts of the seeds are used to treat oligospermia and increase testosterone. Since it is a legume, it has Phyto hemagglutinins and so is contraindicated for patients on blood thinning agents.

Cranberry (*Vaccinium oxycoccus/V. macrocarpon*): Cranberry has been used for preventing urinary tract infections. It has also been used for decreasing the smell of urine in people who are unable to control urination (incontinent).

Cumin (*Cuminum cyminum*): The seeds of this plant are used to spice certain types of foods prominent in the middle east, Indian subcontinent, and South Asian foods. Medicinally the seeds are chewed, or a decoction of seeds are prepared for control of. diarrhea, colic, bowel spasms, and gas. Cumin is also used to increase urine flow to relieve bloating (as a diuretic); to start menstruation, and to increase sexual desire (as an aphrodisiac).

Curcumin – Turmeric (*Curcuma longa*): Curcumin from Turmeric has been claimed to have multiple medicinal properties including anti inflammation, cardio protection, neuro protection and as anti-microbial. Curcumin itself is poorly soluble in water and has poor bioavailability. However, after several years of research supported by the NIH through the National Center for Complementary and Integrative Health no support has been found for curcumin as a medical treatment. Curcumin has been found by the U.S. Food and Drug Administration as a "fake cancer 'cure'". The author of nineteen papers describing the use of curcumin as an anti-cancer agent retracted his papers.

Tannins (a collection of different phytochemicals with common properties and structure) Found in many plant families: They include: 1. Punicalagins found in Pomegranate with potent anti-oxidant properties, 2. Punicacin, a linoleic acid with cardio protective properties, 3. Castalagins – Oak -Quercus spp.; chestnut -Castanea spp.; African birch – Anogeissus leiocarpa. 4. Vescalagins – Oak. 5. Castalins – Broad-leaved paperbark leaves-Melaleuca quinquenervia;, Oak -Quercus spp. 6. Casuarictins – Casuarina and Stachyurus. 7. Grandinins – North American white oak – Quercus alba, European red oak Quercus robur;, broad-leaved paperbark leaves -Melaleuca quinquenervia. 8. Roburin – Oak wood, oak cork. 9. Terflavin – Yellow or chebulic myrobalan – Terminalia chebula. 10. Proanthocyanidins – Maritime pine bark -Pinus pinasters and Pinus maritima, cinnamon – Cinnamomum verum, bilberry-Vaccinium myrtillus, green tea/black tea – Camellia sinensis, Irish oak -Quercus petraea.

Daeba/Dharbha or Kusa grass/Halfa grass, Salt reed grass (*Desmostachya bipinnata*, Syn: *Eragrostis cynosuroides*): Dharbha grass extracts are constituents of medications prescribed by Ayurvedic physicians for treating Kidney stones.

Dandelion (*Taraxacum officinalis*): The above ground parts of this plant are used for making Dandelion tea and powders. Consumption of these products increase urine flow. Side effects include thinning of blood and increase of potassium in the body (hyperkalemia). It reduces absorption of quinolone antibiotics like ciprofloxacin, levoquin and other quinolones.

Deserpidine (Canescine)/Indian snakeroot/*Rauvolfia serpentina*): This alkaloid is an antihypertensive (Blood pressure) drug related to reserpine.

Dill (*Anethum graveolens*): Dill seeds and seed essential oil are used in cooking and as a phytomedicine. The seed is aromatic, carminative, mildly diuretic, galactagogue, stimulant and stomachic. The seed relieves intestinal spasms and griping, helping to settle colic. Chewing the seed improves bad breath. Dill is also a useful addition to cough, cold and flu remedies. It can be used with antispasmodics such as *Viburnum opulus* to relieve period pains. Dill will also help to increase the flow of milk in nursing mothers.

Dimethyltryptamine (DMT): This is a hallucinogenic found in some plants as well as inside the brains, blood, and urine of mammals. Plants that have DMT are various species of *Acacia* (Fabiaceae), *Delosperma* (Aizoaceae), *Desmodium* (Fabaceae), *Lespedeza* (Fabaceae) and Mimosa (few species in family Fabaceae). It is also found in *Tetradium ruticarpum, Euodia leptococca, Pilocarpus organensis, Limonia elephantum*, Several species of *Psychotria*, Dictyoloma incanescens, Dutaillyea drupacea and *D. oreophila, Pandanus* sp, *Virola venosa, Horsfieldia superba, Iryanthera macrophylla, Osteophloem platyspermum, Diplopterys cabrerana, Nectandra megapotamica, Petalostylis labicheoides, P, P. cassinoides, Erythrina flabelliformis, Phyllodium pulchellum, Vepris ampody* and *Zanthoxylum arborescens*.

Dioscmin: This is a bioflavonoid found in orange peels and rind that helps clear up leg sores, spider veins and venous insufficiency.

Dong Quai: Known as 'Female Ginseng" (*Angelica sinensis*): The roots are used to treat pre- and post-menopausal syndrome. It is also believed to stimulate hemoglobin production. This plant is a part of Chinese and East Asian medicine.

Dragons Blood Tree (*Harungana madagascariensis*): The bark, sap and gum are aphrodisiac, astringent, oxytocic, emetic, emmenagogue, expectorant, hemostatic, purgative, styptic and vermifuge. The bark, gum and sap are also effective against the bacterium *Staphyl ococcus*.

Ellagic acid (blackberries, cranberries, grapes, pomegranates, raspberries, strawberries, walnuts, wolfberries, and peach.: Ellagic acid is taken orally for control of bacterial infections and in creams as a skin lightening agent. Ellagic is poorly absorbed.

Emetine – Ipecacuanha/*Carapichea ipecacuanha*: Emetine is a powerful anti-amoebic and antiparasitic agent. It causes nausea and can lead to cardiac failure.

Epena (*Virola sp.*): Snuff powders are used to produce hallucinogenic effects.

Ephedrine/Ephedra/*Ephedra sinica*: It is used to prevent low blood pressure during spinal anesthesia. It has also been used for asthma, narcolepsy, and obesity but has severe side effects causing loss of sleep, appetite, cause stroke and heart attacks.

Ergonovine/Ergometrine – from fungus *Claviceps purpurea*: Claviceps fungus is a contaminating fungus found in some harvested wheat. Ergometrine, also known as ergonovine, is a medication used to cause contractions of the uterus to treat heavy vaginal bleeding after childbirth. Side effect include high blood pressure, vomiting, seizures, headache, and low blood pressure. Other serious side effects include burning of skin (St Anthony's fire) and potentially fatal effects in high doses.

Eyebright (*Euphrasia rostkoviana/E. officinalis*.): All the above ground parts are used in herbal concoctions. Both oral supplementation as well as direct application to eye improves eye strain, inflammation, redness, and dry eyes.

False marijuana (*Zornia latifolia* 'Maconha brava): It is nicknamed *Maconha brava* because locals use it as a cannabis substitute.

Fennel (*Foeniculum vulgare*): Fennel leaves, stem and seeds are used as culinary flavoring agents and in perfumery as well as aroma therapy. The seed extracts and seed oil are used to prepare gripe water to control indigestion, heartburn, intestinal gas, bloating, loss of appetite. and colic. The seed extracts are good for improving and supporting good eye health.

Fenugreek (*Trigonella foenum grecum*): Besides other properties described earlier, for centuries fenugreek has been known as an excellent remedy to increase male libido. These properties are attributed to a high content of diogenin, a natural estrogen.

This herb has been used as a medicine and as a spice for thousands of years in the Indian subcontinent, Middle East, and Asian cuisines. Benefits of fenugreek for diabetes have been proven in both animal and human trials. Side effects of fenugreek may include diarrhea; a maple-like smell to urine, breast milk, and perspiration; and a worsening of asthma.

Fenugreek fiber, Saponins and anti-diabetic 4-hydroxyisoleucine (*Trigonella foenum-grecum*): The seeds contain fiber which can be purified and used in formulations including capsules for helping the body intake of fibers useful for reducing cholesterol and in enhancing regular bowel movements. The saponins in Fenugreek are anti-microbial. The non-essential amino acid 4-Hydroxyisoleucine (4-HIL) is anti-diabetic and works to overcome insulin resistance. This amino acid is the main factor involved in the beneficial use of Fenugreek in traditional medicine. 4-HL also lowers serum triglycerides and lipids.

Flavanols: These are compounds often called Catechins. They include Fisetin (strawberries, mango, persimmon,), Myricetin (nuts, berries, fruits), Quercetin and Rutin. Fisetin is believed to help increase life span. Myricetin has antiviral properties and Quercetin is reported to have unconfirmed anti-cancer properties and is reported to control spread of vitiligo. Rutin is claimed to help control venous insufficiency.

Flaxseed, Linseed, (*Linum usitassimum*): Linseed oil and seeds are rich in α-linolenic acid, an essential fatty acid that appears to be beneficial for the heart diseases, inflammatory bowel disease, arthritis, and other health problems. Several studies suggest that diets rich in omega-3 fatty acids lower BP.

Fo-ti (*Reynoutria multiflora*): Root extracts are prescribed for preventing memory loss and skin diseases.

Fox glove Lady's glove/Hritpatri, Tilapushpi (*Digitalis purpurea*): The leaves have two cardiac glycosides namely digoxin and digoxygenin. Extracts of leaves and pure glycosides are used for treating heart conditions.

Frankincense/Indian frankincense/Shallaki/Salai Guggul (*Boswellia serrata*): This plant is a native of India and Pakistan. Extracts of this plant have several derivatives of Boswellic acid. The plant produces a resin from which an essential oil is prepared. The plant extracts in oil, powder in capsules are used in herbal formulations for control of arthritis and other inflammatory diseases. It is also considered useful for control of diabetes and there are also claims that it is useful for treating cancers.

Frankincense/Shellaki/Olibanum (*Boswellia sacra*): The resin, essential oil, bark, leaf, and root powders are used in treating arthritis and skin diseases. This species is found in Oman, Yemen, and Somalia.

Galangal (Greater Galangal-*Alpinia galanga*, lesser galangal – *Alpinia officinarum*, Chinese ginger – *Boesenbergia rotunda*, black galangal – *Kaempferia galanga*): These species are all members of the Ginger family and the rhizomes are used for culinary and medicinal purposes such as control of nausea and vomiting and other digestive issues.

Gamma oryzanol (Rice bran oil, wheat bran): Rice bran oil rich in this phytochemical is used to lower cholesterol and improve post-menopausal symptoms.

Garlic (*Allium sativum*): Currently, garlic is used as a dietary supplement for many purposes, including high blood cholesterol, high blood pressure, and the common cold, as well as in attempts to prevent cancer and other diseases. Garlic has long been used to help reduce serum cholesterol levels and in general to improve heart health. Garlic controls blood sugar levels also. Fresh garlic powder, and garlic oil are used to flavor foods. Garlic dietary supplements are sold as tablets or capsules. Garlic oil may be used topically (applied to the skin). Side effects include breath and body odor, heartburn, and upset stomach. Taking garlic may increase the risk of bleeding especially if you take an anticoagulant-blood thinner such as warfarin (Coumadin). Garlic bulb extracts have anti-bacterial properties.

Onion, (*Allium cepa*) is also rich in flavins and sulfur compounds. The onion seeds have an oil that can be expressed by cold press. This oil is used in skin and hair care products due to its high essential fatty acid, anti-inflammatory, anti-bacterial, and antioxidant content. As a cleanser, this oil is less drying than normal facial cleansers. It is commonly used in facial creams, hair conditioners, sun care products and massage oils/Both garlic and onion bulb extracts are used in home remedies to control ear infections and earache. The antibacterial activities of the bulb juice are the basis for their use.

Garudakkodi/Eswaramooli/swarmul (*Aristolochia indica* Linn): Antidote for snake bites, Abortifacient, Stimulant, Emmenogogue, Blood purifier. For poisonous bites of snakes, the leaves are ground well and applied or rubbed well over the bitten area. Also 10-20 ml of leaf juice is given with pepper powder internally for 6-7 times daily. The drug is prescribed for the preparation of medicated oil for treating fever with rigor and schizophrenia, and apasmara(epilepsy) in Caraka Samhita. In Sushruta samhita it has been mentioned for the preparation of lepa for Sarpa visha- snake poison.

Genistein – (Dyer's Broom-*Genista tinctoria*, kudzu -*Pueraria lobata*, sohphlang -*Flemingia vestita*, Flemingia -*Flemingia macrophylla*, fava beans, soybeans): Genistein is an important isoflavanoid with multiple medicinal properties. It is an angiogenesis inhibitor and as such helps in cutting off blood supply to cancer cells/tumors. But at the same time, it accelerates spread of estrogen dependent breast cancer cells and acts as an inhibitor of the anticancer drug Tamoxifen. Genistein among other flavonoids is suspected to increase risk of infant leukemia when consumed during pregnancy. It is active in preventing atherosclerosis. It is also a powerful anthelminthic drug.

Ginger (*Zingiber officianalis*): Ginger rhizome powder, extracts, tea, candy, and other forms are excellent aids for control of hyperacidity, flatulence, nausea, and general digestive issues. Ginger (*Zingiber officinale*) is one of the functional foods which has biological compounds including gingerol, shogaol, paradol and zingerone. Ginger has been proposed to have anti-cancer, anti-thrombotic, anti-inflammatory, anti-arthritic, hypolipidemic, hypoglycemic and analgesic properties. Ginger, Dandelion tea, Chamomile tea, Valerian root extracts, St. John's Wort, Echinaceae, Ginkgo, Kava and stinging nettle have been suggested as potential herbals to help improve symptoms of Multiple sclerosis.

Gingerol – Ginger (*Zingiber officinale*). Gingerols are phenolic compounds in ginger roots. These compounds have antioxidant and anti-inflammatory activities. They are effective in controlling nausea and upset stomach. Many studies show that they are effective in controlling colitis.

Gingko (*Ginkgoa biloba*): Ginkgo leaf extracts, tea and tinctures are prescribed by practitioners of Chinese traditional medicine for improving cognitive functions. There is some evidence that Gingko extracts can control tinnitus. It also helps good blood circulation.

Ginseng (*Panax ginseng*): Ginseng has been used as a traditional medicine for more than 2,000 years. Studies suggest that both Asian and American ginseng may help lower blood sugar in people with diabetes. Ginseng roots are believed to prevent dementia. It is said to improve testosterone levels and help in muscle building. It is an important herb in the Chinese herbal pharmacopeia.

Panax Ginseng is a common ingredient in many Chinese and oriental herbal preparations for correcting erectile dysfunction, premature ejaculation, and other sex related functions.

Ingestion of Ginseng extracts improves memory in older populations. It is said to improve reaction times, arithmetic skills, and memory in general. Ginseng is also prescribed to correct erectile dysfunction, improve sexual desires, and improve lung functions.

Glucosinolates (Broccoli, Kale, Mustard greens, Rape): Glucosinolates are under investigation as anticancer agents.

Goat Weed (*Epimedium* sp.): horny goat weed is described in Chinese medicine as a natural alternative to drugs for erectile dysfunction (ED). Horny goat weed may result in irregular heartbeat, low blood pressure, nosebleeds, and mood changes. High doses have been linked to spasms and respiratory failure.

Goji Berries/Wolf Berries (*Lycium barbarum*): The root bark and berries are used as medicine. Goji is used for many conditions including diabetes, weight loss, improving quality of life, and as a tonic. As per Chinese medicine, Goji berries control cholesterol and helps avert development of Diabetes.

Gou teng/cat's claw (*Uncaria rhynchophylla*): Gou teng is a part of TCM used to treat Parkinson's disease. It is believed to have anti spasmodic properties.

Grape seed (*Vitis vinifera*): The juice from stem is used as eye drops. Fruits hold many chemicals including resveratrol, many different anthocyanins, other antioxidants, phenolics and sugar. Various studies have shown that resveratrol helps to keep good cardiac health. Apart from its use in wine making, grape juice extracts are used in Ayurvedic medicines and tonics. Grape seed oil is rich in polyunsaturated fats and helps in reducing total cholesterol and increasing HDL. The seed oil has unsaturated fatty acids that help to lower LDL cholesterol.

Graviola (*Annona muricate*): In animal models and cell lines, Graviola extracts showed anti-Cancer, anti-inflammatory, analgesic, antidiabetic, antiulcer, and antiviral effects. The leaf extracts also have antimicrobial activities. However, Graviola products have not been studied in cancer patients. Alkaloids present in Graviola caused movement disorders and myeloneuropathy with symptoms mimicking Parkinson's disease in vitro.

Green Tea (Camellia sinensis): Green tea is rich in flavanoids and antioxidants and so daily intake of green tea infusions are cardio protective and help to support good health.

Guduchi/Giloy/Madhuparni/(*Tinospora cordifolia*): This is an important herbal in the Ayurvedic system of herbal medicine. The leaf powders are prescribed for improving cardiac health, stimulate digestion and relieve constipation. It is a general tonic.

Guggul/Indian bdellium tree/Mukul/(*Commiphora wightii*): studies in India show that it has powerful cholesterol lowering properties. The active ingredient in Gugul lipid is gugulsterone.

Happy tree/tree of Life Xi Shu (*Campotheca acuminata*): The bark and stems of *Camptotheca acuminata* have the alkaloid camptothecin. Several chemical derivatives of camptothecin are used as drugs for cancer treatment, topotecan for ovarian, lung and other cancers, rubitecan for pancreatic cancer, Irinotecan, sold under the brand name Camptosar to treat colon cancer, and small cell lung cancer. Another analog, Belotecan is used for Small Cell Lung Cancer and Ovarian Cancer, approved in South Korea under the trade name Camtobell®.

Hardy Fuchsia/Chiko/Tilco (*Fuchsia magellanica*): This plant is native to Southern Argentina and Chile. Infusion of the leaf extract reduces body temperature and acts as a diuretic, and lowers BP.

Hawaiian Baby Woodrose (*Argyreia nervosa*): Seeds are rich in Lysergic acid-LSA (also known as d-lysergic acid amide, d-lysergamide, ergine, and LA-111). This is a powerful hallucinogen.

Hawthorn/Mayflower (*Crataegus laevigata*): Hawthorn can improve heart health. The leaves and flowers are steeped in water overnight and drunk as a tea or a tincture of the leaves and flowers are prepared for treatment.

Henna/Mehndi: (*Lawsonia inermis*): Henna leaf extracts are used for various cosmetic conditions. Henna is applied directly to the affected area for dandruff, eczema, scabies, fungal infections, and wounds. Henna extracts are also used to treat stomach and intestinal ulcers.

Hesperidin, (Citrus fruit): The flavanone glycoside bioflavonoid Hesperidin that is found in citrus fruits, has anti-inflammatory and cardiovascular protective properties. It improves peripheral circulation including venous insufficiency, correcting spider veins and other leg circulation problems. Being an antioxidant, it is also reported to help prevent or reduce the symptoms of neurodegenerative diseases such as Parkinson's, Alzheimer's, Huntington's, and multiple sclerosis.

Hibiscus (*Hibiscus subdariffa, Hibiscus Roselle, H. rosasinensis, H. mutablis, H. tiliaceus, H. hirta, H. hispidissimus, H. arnottianus*). Studies conducted by USDA scientists show that Roselle tea *H. Subdariffa* consumption lowers blood pressure due to inhibition of Angiotensin. Hibiscus tea is a common drink in the middle east. The shoe flower (*Hibiscus rosasinensis*) extracts mixed in oil are applied on the skin for general skin health and in oil and shampoos for good scalp health. Flowers steeped in hot water and then chilled are consumed in tea like preparations.

Holy basil, (*Ocimum sanctum*): This herb is commonly used in India as a traditional medicine for diabetes. Studies in animals suggest that holy basil may increase the secretion of insulin. A controlled trial of holy basil in people with type 2 diabetes showed a positive effect on fasting blood sugar and on blood sugar following a meal.

Hoodia/Kalahari Cactus (*Hoodia gordonii*): Hoodia dietary supplements are used as an appetite suppressant for weight loss. Hoodia is available as liquids, powders, tablets, and capsules. Some hoodia products also have other herbs or minerals, such as green tea or chromium. Although Hoodia is a part of some anti diabetic and weight loss supplements the safety of this product has not been fully evaluated. In one study, participants taking hoodia had more side effects than those taking placebos Side effects included nausea, vomiting, dizziness, and odd skin sensations and increases in blood pressure.

Hops (*Humulus lupulus*): The female and male cones of the plants have many phytochemicals and extracts are commonly used in the beer industry. These also have mild sedative properties.

Horse chestnut (*Aesculus hippocastanum*): Seeds have about 20% Aescin that is efficacious in treating chronic venous insufficiency, varicose veins, and hemorrhoids. Both whole seed standardized horse chestnut powder having no more than 100-150 mg of aescin and a gel having 2% aesin are available commercially for treatment of swelling of legs due to chronic venous insufficiency.

Hydrocurcumins (*Curcuma longa*): This is a colorless hydrogenated product derived from the yellow curcumins and function as antioxidant compounds. Because of the lack of yellow color, it is used in cosmetics that must be free of the yellow color but have antioxidant properties.

Hydroxycinnamic acids (HCs) (coumaric acid, ferulic acid, sinapic acid, caffeic acid, chlorogenic acid, Caffeic acid rosmarinic acid): All are phenolic compounds found in fruits, vegetables, and beverages-coffee, tea, wine. Several studies have reported that HCs and their derivatives function as antioxidants and protect biologically important molecules from oxidation.

India drumstick (*Moringa Indigofera*): Ayurvedic practitioners prescribe drumstick fruits as an aphrodisiac. The plant leaves and fruits have other useful medicinal properties. The leaves are rich in vitamin C.

Indian Pennywort/Gotu Kola/Manduukaparni (*Centella asiatica*): Extracts of the leaves are considered cardioprotective and some studies show that it has anti-ageing properties.

Indian snakeroot/Sarpagandhi (*Rauvolfia serpentina*): Roots and other plant parts produce several indole alkaloids of which reserpine is important because of its anti-hypertensive and anti-psychotic effects. The active compound reserpine was marketed under the name's reserpine, serpacil, Raudixin for controlling hypertension. However, it is no longer prescribed because of its side effects.

Indian White Cedar/Bombay white cedar (*Disoxylum binectariferum*): Crude extracts of the tree were found to be highly effective against ovarian and breast cancer. Rohitukine, a chromane alkaloid, is a precursor of flavopiridol, a promising anti-cancer compound. Flavopiridol has been shown to be a potent cyclin-dependent kinase inhibitor. Alvocidib (INN; also known as Flavopiridol) is a flavonoid alkaloid CDK9 kinase inhibitor under clinical development for the treatment of acute myeloid leukemia, by Tolero Pharmaceuticals, Inc. It has been studied also for the treatment of arthritis. The FDA has granted orphan drug designation to Alvocidib (flavopiridol) to treat patients with acute myeloid leukemia. The Orphan Drug designation—to encourage development of drugs in the diagnosis, prevention, or treatment of a medical condition affecting fewer than 200,000 people in the U.S.—grants a product market exclusivity for a seven-year period if the sponsor complies with certain FDA specifications, as well as tax credits and prescription drug user fee waivers. Flavopiridol and related compounds are also being evaluated for treatment of Glioblastoma -a deadly form of brain tumor.

Indian Sarsaparilla/Sariva/Nannari/Anantamula, (*Hemidesmus indicus*): The roots of this plant have medicinal qualities such as diabetic control, urinary tract health by improving urine flow (Mehanashana), control of urine smell (Durgandha Nashana) and sperm production.

Indole-3-carbinol – (Broccoli, Cabbage, Cauliflower and Brussels sprouts): This indole phytochemical is reported to help prevent/treat breast, cervical cancer as well as Lupus.

Ingenol mebutate (Milk weed): This phytochemical is approved for use in creams and topical applications for actinic keratosis (warty overgrowths of skin) by the FDA, USA. But the cream can cause severe allergies and even skin cancers. The drug has been withdrawn from the European Union.

Insulin Plant/Fiery Costus/Spiral Flag (*Chamaecostus cuspidatus*): it has been proven to have various pharmacological activities like hypolipidemic, diuretic, antioxidant, anti-microbial, anti-cancerous. It is used in India to control diabetes, and it is known that diabetic people eat one leaf daily to keep their blood glucose low. Plant extracts also have diuretic properties.

Inulins (Fruits, vegetables, and herbs, including Wheat, Onions, Bananas, Leeks, Artichokes, and Asparagus): Inulins help to reduce cholesterol and triglycerides and help control diabetes.

Ipecac (*Carapichea ipecacuanha*): Ipecac root extract powders or syrup are used as anti-nausea agents to induce vomits to expel poisons. But it is also anti amoebic in nature.

Ivy Gourd/Kova Kai/Donda Kaya/Tindora (*Coccinia grandis*): Fruits of this plant are reported to have antioxidant, antihypoglycemic, and immune system modulator activities. Root extracts are used to control osteoarthritis in Bangladesh.

Jack Fruit/Jaca/Chakka/Pala pazham (*Artocarpus heterophyllum*/*A. integrifolia*/*A. braziliensis*: This member of the Moraceae family is closely related to bread fruit plant. The plant produces compound fruits where the fleshy individual fruits as well the seeds are edible. The seeds are rich in fiber and complex carbohydrates and protein and hence the seed meal could be a good source of carbohydrates as a substitute or additive to cereals. The fleshy unripe and ripe edible fruits are also rich in carbohydrates, proteins and fiber and have a low glycemic index. The fruit and seed flour are prescribed for easy bowel movement. The fruits and seeds are considered as good food for diabetic control. The fruits are also reported to be aphrodisiacs.

Jiaogulan (*Gynostemma pentaphyllum*): The plant has been used in Chinese traditional medicine for its adaptogen and antioxidant properties that are believed to increase longevity. Leaf extracts of this Southern Ginseng is used in Traditional Chinese Medicine by mouth for treating high cholesterol, diabetes, liver disease and obesity.

Jimson weed (*Datura stramonium*), Deadly Nightshade (*Atropa belladonna*) and Mandrake (*Mandragora officinarum*): All these plants have the active hallucinogenic atropine and scopolmine. Ingesting any of these plants will cause violent hallucinations, seizures and in many cases death.

Karangin (*Pongamia glabra*): The seeds and oil from this plant have insecticidal properties and is used as a vermifuge in humans and animals. It is also an organic crop insecticide.

Karpooravalli (*Coleus forskohlii*): This vine leaves smell of camphor. Coleonol, a diterpene, isolated from *C. forskohlii*, have been shown to lower the BP of anesthetized cats and rats.

Kava (*Piper methysticum*): Kava root extracts in water induce sleep and has some anesthetic properties. However, solvent extracts can be hepato (Liver) toxic.

Khat or qat (*Catha edulis*): Khat has the alkaloid cathinone, a stimulant, which is said to cause excitement, loss of appetite, and euphoria.

Kidney Stone inducers: Consuming the plant products in large amounts of certain plants described in Chapter 8 can induce kidney stones mainly because they are rich in oxalic acid which upon ingestion interacts with calcium to for calcium oxalate crystals that form kidney stones.

Kizha Nelli/Tamalaki/Stonebreaker (*Phyllanthus niruri/P. amarus*): This relative of P. emblica is used to break up and expel kidney stones and is also prescribed for treating liver cirrhosis and other liver ailments.

Kurchi/Tellicherry bark (*Holarrhena Antidysenterica syn H. pubescens*): In Indian traditional medicine, *H. Antidysenterica* is popularly used as a medication for dysentery, diarrhoea and intestinal worms. Plant parts such as bark are used as anti-microbial, anti-inflammatory, analgesics, anti-amoebiasis, chronic bronchitis, locally for boils, ulcers. A decoction of the bark is used in bleeding and piles.

Kutki/Indian Gentian (*Picrorhiza kurroa*): Kutki is a bitter ayurvedic herb whose root extracts are prescribed for treating liver conditions. It is also prescribed for treating psoriasis and vitiligo.

Lavender (*Lavendula angustifolia*): Lavender flower extracts and essential oil are used in aromatherapy and in cosmetics and other perfumeries.

Licorice (*Glycyrrhiza glabra*): Licorice root extract and root oil are effective for control of acid reflux but have dangerous side effects in high doses. It can cause heart arrhythmia, inducing high blood pressure, edema, and potassium deficiency.

Lily-of-the-valley (*Convallara majalis*): It is considered by herbalists as an adjunctive or standalone therapy for the treatment of cardiac disorders such as arrhythmia, mitral valve prolapses and shortness of breath.

Limonene – Citrus: Limonene is common as a dietary supplement and as a fragrance ingredient for cosmetics products. It has anti-bacterial properties.

Lobeline/Indian tobacco – *Lobelia inflata*/Devil's tobacco – *Lobelia tupa*/star of Bethlehem – *Hippobroma longiflora*/cardinal flower – *Lobelia cardinalis*/, great blue lobelia – *Lobelia siphilitica*: Lobeline has been sold, in tablet form, for use as a smoking cessation aid and for recovery from addictions. But clinical evidence is mixed.

Long pepper (*Fructus piperis longi/Piper longum*): The seeds and seed extracts are components of various Ayurvedic preparations. They stimulate the appetite, enhance absorption of food, and have antimicrobial activities.

Lycopene (*Lycopersicon esculentum*) Tomato: The tomato fruits have the phytochemical lycopene. Anecdotal and experimental reports suggest that it is useful as an agent for control of benign prostate hyperplasia. (Prostate enlargement).

Maca/Peruvian Ginseng (*Lepidium meyenii*): The underground parts of the plant are prescribed in Peruvian Inca medicine as an aphrodisiac to increase libido, impotence and increase fertility and control inflammations.

Madagascar periwinkle/Vinca/Nithya Kalyani (*Catharanthus roseus*): Madagascar periwinkle had been used for treating diabetes mellitus, high blood pressure and infection. About 70 different indole alkaloids are found in the milky sap from the stems of periwinkle. Two of the common anti-cancer drugs which are derived from this plant are vincristine and vinblastine. Vincristine is used in the chemotherapeutic regime for Hodgkin's lymphoma while vinblastine is used for childhood leukemia.

The Main side effects of these drugs are peripheral neuropathy, hair loss, hyponatremia, and constipation.

Malabar Plum/Java Plum/Jamun/Naval pazham (*Euginea jambolena/ Syzygium cumini*): The fruit juices of the plant have high levels of flavonoids and are prescribed for treating Diabetes, Asthma, constipation. Diarrhea, Gas (flatulence), Skin ulcers, and Swelling (inflammation) of the main airways in the lung (bronchitis).

Malabar tamarind/Kudam Puli/Vrkshaamla (*Garcinia gummi gutta syn G. gambojia*): The fruits of this tree have high levels of Hydroxy citric acid which is popular as a weight reducing agent. However, although it did reduce fats in rat experiments it has had no significant effects in reducing fat in humans. *Garcinia cambogia* has anti-inflammatory properties and it is especially useful for people suffering from inflammatory bowel diseases. It also controls gastric ulcers.

Male fern (*Dryopteris filixmas*): Leaf extracts are used as vermifuges since Roman times. But these extracts are toxic and hence not recommended in current herbal medicine.

Mama Cadela/Sweet Cotton (*Brosimum gaudichaudii*): Root extracts and leaves have certain furanocoumarins that delay or prevent the destruction of melanocytes seen in vitiligo/leukoderma. In view of this root and leaf decoctions are applied to the skin patches showing vitiligo symptoms.

Mango Ginger (*Curcuma amada*): This relative of Ginger and Turmeric are used as appetite enhancers and pickles. But medicinally, extracts have anti-inflammatory, anti-allergy, analgesic and anti-bacterial activities. It has a mango aroma flavor.

Mangosteen/Purple mangosteen (*Garcinia mangostana*): Mangosteen fruit juice and dried fruits are prescribed for reducing weight and treating gum disease. The fruit juice is consumed to relieve muscle weakness after exercise.

Maritime Pine *(Pinus pinaster):* Pycnogenol is an extract from French maritime pine bark. It is an herbal treatment for venous insufficiency and other vascular conditions.

Marshmallow *(Althaea officinalis)*: Marshmallow leaf and root extracts are commonly ingested by mouth to treat stomach ulcers, diarrhea, constipation, swelling of the stomach lining, and pain and swelling of the mucous membranes that line the respiratory tract.

Matcha Green Tea (*Camellia sinensis*): Green tea is richer in antioxidants compared to other forms of tea. Tea is composed of polyphenols, caffeine, minerals, and trace amounts of vitamins, amino acids, and carbohydrates. Green tea has epigallocatechin-3-gallate (EGCG) which is known to control development of prostate cancer and promote prostate health. Observational studies have also shown that drinking green tea is associated with a reduced risk of heart disease and stroke. The phytochemicals present in green tea are known to stimulate the central nervous system and support overall health in humans. Green tea controls photoaging thereby protecting the skin. It is also described as a stress reliever and immune booster.

Mayapple/American mandrake/wild mandrake/and ground lemon (*Podophyllum peltatum, P. hexandrum*): Podophyllum has been used topically in the treatment of genital warts and hairy leukoplakia. Semisynthetic derivatives of podophyllotoxin are used to treat rheumatoid arthritis and various cancers, including refractory testicular tumors and small-cell lung cancer. The root and plant have valuable constituents such as Quercetin, Kaempferol, Podophyllin, Isorhamnetin, Gallic-acid, Berberine, Alpha-peltatin, that are being studied for their healing, anticancer and other properties. There are a few serious side effects.

Mescaline: Peyote cactus (*Lophophora williamsii*): The active compound Mescaline is a hallucinogenic compound. Mescaline is also found in *Opuntia acanthocarpa, Opuntia echinocarpa, Opuntia cylindrica* and *Opuntia basilaris*. It is traditionally used by Native Americans in spiritual rites central to the Native American Church. Though it has been suggested to be an effective treatment for depression and alcoholism, it still is a Schedule I substance which limits its usage in medical research.

Mexican poppy, (*Argemone Mexicana*): Used by Chinese residents of Mexico during the early 20th century as a legal substitute for opium and currently smoked as a marijuana substitute.

Milk Thistle (*Silybum marianum*): This flowering herb is found around the Mediterranean Sea. It has been used for its medicinal properties for thousands of years. It is sometimes known by the name of its active part, silymarin. Milk thistle may reduce insulin resistance in people with type 2 diabetes who also have liver disease. Milk thistle is consumed by mouth most often for liver disorders, including liver damage caused by chemicals, alcohol, and chemotherapy, as well as liver damage caused by Amanita mushroom poisoning, nonalcoholic fatty liver disease, chronic inflammatory liver disease, cirrhosis of the liver, and chronic hepatitis.

Miraculin (Sapotaceae, *Synsepalum dulcificum*): The berries of this tree have glycoproteins that are taste modifiers so that eating the fruit causes the taste buds to think that even sour and bitter foods are sweet. It is particularly useful for patients who are on chemotherapy which gives a metallic taste. Eating the fruit or extracts of the fruit enables patients to taste good foods thus enriching their nutritive status.

Morning glory (*Ipomoea tricolor, I.* violacea): The seeds of the Morning glory plant have lysergic acid hallucinogens. Many of these plants were used as part of the shamanistic rituals of Native Americans; several are still used as recreational drugs.

Morphine/Opium poppy (*Papaver somniferum*): Morphine is a pain medication of the opiate group. It acts directly on the central nervous system (CNS) to decrease the feeling of pain. Modified morphine known as lidocaine is used as a local anesthetic as an injectable or in topical creams.

Mu tong (*Aristolochia manshuriensis*). This Chinese plant is being used as a diuretic and anti-inflammatory for the treatment of edema and rheumatic pain. One component Magnoflorine has been found to possess hypotensive properties.

Myrrha/African Myrrha (*Commiphora Myrrha*): The resin from this Arabian Peninsula and North Africa tree is used as an incense with disinfection properties. The resin and resin oil are part of both Ayurvedic, Unani/Arabian and Chinese herbal traditional medicine. Myrrha is used in mixtures with Frankincense to treat pain due to inflammation. It is also considered to be an Emmenogogue and improve circulation.

Naringenin – (Grapefruits, oranges, mandarin oranges, and tomatoes (peel): It has Antibacterial, antifungal, and antiviral properties. It is effective against *Staphylococcus aureus, Bacillus subtilis, Micrococcus luteus, Escherichia coli, Actinomyces naeslundii, Prorphyromonas gingivalis,* several species of Candida and *Helicobacter pylori.* Naringenin also inhibits Hepatitis C, poloiviris and herpes viruses in cell cultures. It has significant antioxidant properties as well as cytotoxicity towards breast cancer in cell cultures.

Naringin (Citrus fruits): This is another flavanoid glycoside found in citrus fruits. Its role is not clear.

Narkia tree/Kalagaura/Arali/peenari/Dhurvasane mara (*Nathopodites foetida/Mappia foetida*): *Mappia foetida* is grown in India. It has anticancer and antiviral properties. The alkaloids of Narkia, soluble in water, are present in all the parts of the plant, and are the precursors of camptothecin and of 9-methoxy-camptothecin, which are alkaloids known to have pharmacodynamic properties but also to be insoluble in water.

Neem (Azadirachta indica/*Melia azadirachta*): Neem leaf powers, gum and seed oil are powerful insecticides and are used as vermifuges. The leaf paste mixed with turmeric is used as a treatment for control of eczema, psoriasis and skin infections and as antiviral agents. Neem leaf extracts and neem seed oil are used in anti-lice preparations. The azadiractin in neem has anti-microbial properties and hence is a part in some soaps and shampoos. Neem oil and extracts are also good anti parasitic agents to clear intestinal worms. In India it was a common practice to apply neem paste topically on skin of patients with smallpox, chicken pox and measles virus diseases. It was also a practice to spread Neem leaves in the beds of these viral infected patients.

Nicotine (Wild tobacco -Nicotiana rustica, tobacco plant -Nicotiana tabacum, pituri -*Duboisia hopwoodii*, common milkweed -*Asclepias syriaca*): Nicotine is a widely used stimulant and is used for relief of smoking withdrawal symptoms. It is addictive.

Nut Grass/Musta/Muthanga/Nagarmotha (*Cyperus rotundus* Linn.): The underground rhizome extracts are used for treating menstrual issues such as amenorrhea. It also stimulates urination. Extracts are also components of some Ayurvedic oils used for improving skin ailments.

Nymphaea caerulea (Blue lotus or lily); *Nelumbo nucifera,* (the Sacred Lotus): Both *Nymphaea caerulea* and *Nelumbo nucifera* which have the alkaloids nuciferine and aporphine have been used as a sacrament in ancient Egypt, India and some South American cultures. The flower extracts of both plants are used in aromatherapy. Nuciferine crosses the blood brain barrier and some studies show that is effective in slowing down the progression of glioblastoma brain cancer. It is also reported to be useful in treating erectile dysfunction. Experiments with diabetic animal models suggest that it can heal liver damage.

Oats (*Avena sativa*): A diet containing soluble fiber-rich whole oats can significantly reduce the need for antihyp ertensive medication and improve BP control. Oats also exhibits glucose control.

Ochrosia (*Ochrosia oppositifolia*): Leaves and flower extracts have abortifacient properties.

Okra/Lady's finger *(Hibiscus esculentes)*: Okra is known to have a positive effect on blood sugar control, among many other health benefits. Evidence of the antidiabetic properties of okra and its oligosaccharides has been reported through several studies.

Oleander (*Nerium indicum; N. odorum, Thevetia peruviana*): Oleander seed extracts are used to induce abortions. But at the same time, oleander glycosides are used to treat heart conditions, asthma, epilepsy and even cancer.

Oleanolic acid – Olive oil – (*Olea europaea*): Oleanolic acid is non-toxic, hepatoprotective, and shows antitumor and antiviral properties. It may contribute to the health benefits of omega 3 oil in olive oil.

Opium poppy/Morphine/Opiates (*Papaver somniferum*): Resin from opium poppy is rich in morphine which is an immensely powerful pain killer. Opium as well as all opiates are habit forming and should be used with caution.

Oregano (*Oregano vulgare*): The leaves have an essential oil which has anti-microbial properties. Fresh and dried Oregano leaves are used to spice up foods.

Other common Sleep aid herbs: Other natural herbs that have been used as herbal sleep aids include: Cowslip – (*Primula veris*), Great Mullein – (*Verbascum thapsus*), Mugwort – (*Artemisia vulgari*), Bugleweed – (*Lycopus europaeus*), Jamaica Dogwood – (Piscidia erythrina), Mistletoe – (*Viscum album*), Lemon Grass – (*Cymbopogon citratus*), Lavender – (*Lavandula angustifolia*), Vervain – (*Verbena officinalis*), Heather – (Calluna vulgaris), American Ginseng – (Panax quinquefolius), Common Mallow – (*Malva sylvestris*), Sweet Basil – (*Ocimum basilicum*).

Other herbs for skin: Extracts of Dandelion flowers and leaves, Elderberry flowers, Borage, Chamomile, Parsley, Rosemary, Thyme, Yarrow (*Achillea millefolium*), Marshmallow (*Althaea officinalis*), carrot seed oil, Rose petal oil, Plantain leaf extracts and St John's wort aerial parts are all components of various skin and hair supplements. Mucilaginous extracts from English plantain (*Plantago lanceolate*), fenugreek (*Trigonella foenum-grecum*), mullein (*Verbascum thapsus*), slippery elm (*Ulmus fulva*), and flax (*Linum usitatissi-mum*) contain mucilage's, which act as emollients on and soothe the skin.

Other plants that might have a role in Cancer therapy: *Curcuma longa* (Turmeric), *Eugenia jambolana* (Jambul fruit), *euphorbia peplus, lavandula angustifolia, Maytenus ovatus, Mentha viridis Ocimum sanctum, patrinia villosa, perilla frutescens, physalis angulate, salix alba, salvia officinalis, scutelaria baicalensis, scutellaria baicalensis, solanum nigrum, perilla frutescens* and *Mentha viridis*.

Pacific yew (Taxus brevifolia, Canadian yew (*Taxus canadiensis*), European Yew (T. baccata): In 1963, the National Cancer Institute found that extracts of *Taxus brevifolia* (Pacific yew) bark were active in vitro against cancer cells. Crude extracts of the plant were effective against various types of cancer cell lines. The active taxane diterpenoid paclitaxel ended up as an FDA approved chemotherapy drug named Taxol®.

Pacific yew/Anglo Japanese yew/English Yew (*cephalotaxus harringtonia*): Homoharringtonine, an alkaloid isolated from *Cephalotaxus harringtonia* is an approved treatment of adult patients with chronic myeloid leukemia.

Papaya (*Carica papaya*): Carica papaya fruits are good for digestive diseases for improving digestion of proteins. The latex is believed to induce abortions. Leaf infusions are reported to be good for diabetic control. the aqueous extract of C. papaya exerted a hypoglycemic and antioxidant effect. It also improved the lipid profile in diabetic rats. In addition, the leaf extract positively affected integrity and function of both liver and pancreas. Papaya fruits are popularly believed to induce abortions. 10-15 ml of latex of raw fruit is given orally once a day for 3 days for inducing abortions.

Passionflower/May pops (*Passiflora incarnata*): Passionflower tea is also a part of many sleep preparations.

Pau d'arco/Lapacho/Trumpet tree (*Tabuia avellanedae/Handroanthus impetiginosus*): It is used in Brazilian herbal medicine for many conditions including cancer, ulcers, diabetes, candida, rheumatism, arthritis, prostatitis, dysentery, stomatitis, and boils. In North American herbal medicine, pau d'arco is used as an analgesic, antioxidant, antiparasitic, antimicrobial, antifungal, antiviral, antibacterial, anti-inflammatory, and laxative. Pau d'arco also is employed in herbal medicine systems in the United States for lupus, diabetes, ulcers, leukemia, allergies, liver disease, Hodgkin's disease, osteomyelitis, Parkinson's disease, and psoriasis. The American cancer society does not endorse the anti-cancer claims for this plant.

Pepitas/Pumpkin seeds (*Cucurbita pepo*): Pumpkin seed oil has aphrodisiac properties. It is also considered to be cardioprotective. Pumpkin seed oil and extracts. The seeds are rich in beta sitasterol which is reported to help in reducing prostate inflammation.

Peppermint (*Mentha ×piperata*): Peppermint oil extracted from leaves is used for treating irritable bowel syndrome or IBS, indigestion (dyspepsia) and spasms in the bowel. Peppermint oil is also used in aromatherapy.

Perillyl alcohol: It is found in the essential oils of plants like lavender, lemongrass, sage, peppermint, mint. Perillyl alcohol has shown some antitumor activity in laboratory and animal studies.

Phenolic compounds: Phenolics like Apiole, Carvacrol have some value in health. Apiole oil is an abortifacient and influences menstrual problems. It is toxic in high concentration. Carvacrol found in Thyme, oregano, Marjoram is effective against many bacterial strains.

Phytic acid (wheat, many grains): This inositol polyphosphate found in wheat and many grains binds/chelates many essential minerals like iron making them unavailable to the body. Those who are iron deficient (most vegans and vegetarians) should not take iron supplements along with meals having phytates.

Pineapple/Annasi/Kannara (*Ananas comosus* L.): Ripened fruit is used to induce abortion.

Piperine (Black pepper -Piper nigrum, long pepper -Piper longum, Balinese long pepper-Piper retrofractum): Piperine is said to increase bioavailability of food and dietary supplements, such as curcumin from turmeric. It has a strong pungent taste and is also used in pepper sprays.

Poinciana/peacock flower,/red bird of paradise/Mexican bird of paradise/dwarf Poinciana/pride of Barbados,/flos pavonis, and flamboyant-de-jardin/Rajamally(-Tamil-India:(*Caesalpiniapulcherrima* (L.): Bark juice (2 ml) is administered orally on empty stomach for the first three months of pregnancy in order to induce abortions.

Poly herbal formulation/Nilavembu Kudineer: This is an herbal polyherbal formulation that is used for treating the severe break bone symptoms, fatigue, headache, and other symptoms of Dengue virus infections. It is also claimed to boost the immune system and increase platelet count. Siddha medicine practitioners have clinically evaluated this formulation in patients with Dengue fever and have reported positive effects. The herbal constituents are described in Chapter 12.

Pomegranate (*Punica granatum*): Powerful antioxidant chemicals in pomegranate fruit and juice may help reverse atherosclerosis and lower blood pressure. The fruit juice is also a good diuretic.

Prickly Pear Cactus (*Opuntia spp.*): The Cladode shows promise in potential glucose-lowering effects. But, currently, there is a lack of evidence to support the recommendation of using Opuntia spp. fruit products as an alternative or complementary therapy in the reduction of risk or management of Type 2 Diabetes Mellitus.

Protodioscin: This steroidal saponin compound found in *Tribulus terrestris*, *Trigonella foenum-grecum* (Fenugreek) and *Dioscorea* (Yam) stimulates testosterone production. It is reported to be beneficial for correcting erectile dysfunction.

Psilocybin psychedelic mushrooms native to Mexico, Central America, and the US. Psilocybin can produce hallucinations, an inability to differentiate reality from fantasy, panic attacks, and psychosis if consumed in large doses. It is also known as Magic Mushrooms and Shrooms.

Pterostilbene: (heart wood from *Pterocarpus marsupium* (Indian Kino Tree). Pterostilbene is a stilbene chemically related to resveratrol and is found in blueberries and grapes. it is thought to show anti-cancer, anti-hypercholesterolemia, anti-hypertriglyceridemia properties, as well as reverse cognitive decline. It is believed that the compound also has anti-diabetic properties.

Punarnavaa (*Boerhavia diffusa*): Although this plant is considered to have beneficial effects on the Kidneys, plant extracts of this plant are mainly considered to be re-juvenating agents and is a component of some Ayurvedic tonics like Chyavanaprash.

Pycnogenol (European Pine bark): Pycnogenol helps to control Asthma and allergies. It is also used to improve arterial circulation and prevent blood clots.

Quinine tree/Jesuits bark (*Cinchona officianalis*): Cinchona bark powder and extracts have anti-malarial properties and were the original ingredient of tonic water. Quinine is also used for control of leg cramps.

Quinine – (Cinchona tree -*Cinchona officinalis*): Quinine alkaloid from bark of Cinchona trees is used to control Malaria and for control of leg cramps. Side effects are headache, ringing in the ears, deafness, irregular heartbeats, trouble seeing, and sweating.

Radish seeds (*Raphanus sativus*): Herbalists recommend application of a paste of powdered radish seeds to the white skin patches of patients with vitiligo. The paste can also be incorporated in skin creams along with other ingredients.

Ramachempu (*Rhynchosia rufescens*): Leaf decoction is administered for inducing abortions among the Mannan tribes in Kerala, India.

Reserpine – (Indian snakeroot -*Rauwolfia serpentina*, devil's pepper-*Rauvolfia vomitoria*). Reserpine alkaloid was one of the first hypertensives to be marketed for blood pressure control but is no longer used since better medicines are now available.

Resveratrol: skin of grapes, blueberries, raspberries, mulberries, peanuts, and other plants: Many animal studies have shown that resveratrol reduces age related symptoms in animal studies. However, human studies do not fully back this.

River Lily, Swamp Lily (*Crinum glaucum*): The extracts of this plant are used traditionally in Western Nigeria for treatment of asthma. It was found to decrease both systolic and diastolic pressures.

Rosemary (*Rosmarinus officinalis*): Rosemary oil is an aromatic agent for flavoring and aromatherapy. It is regarded as being able to stimulate hair growth. It is also regarded as an agent that promotes abortions, but the evidence is very scanty.

Rosmarinic acid – (Basil, holy basil, marjoram, rosemary, lemon balm, sage, selfheal, velvetleaf soldierbush-*Heliotropium foertherianum*). Both oral preparations as well as topicals having this acid are good antioxidants.

Russian Tarragon (*Artemisia dracunculus L.*): Recent studies reported antidiabetic and hypoglycemic activities. An ethanolic extract of RT was able to reduce blood glucose concentration in rodents. The aqueous extract of RT is a promising and safe ingredient for consideration in the development of functional foods or dietary and sports supplements with anti-hyperglycemic activity.

Saffron (*Crocus sativus*): Saffron supplementation is being promoted for treating ocular neurodegenerative pathologies, such as diabetic retinopathy, retinitis pigmentosa, age-related macular degeneration and glaucoma, among others, and shows beneficial effects. Clinical evidence is inconclusive but current experiments support further investigations. Stigmatic powders from Saffron added to milk or other beverages and food are believed to help in keeping and improving cognition.

Salacia/Chundan, Kothala Himbutu, Ponkoranti (*Salacia reticulata/ S. oblonga*): The leaf and bark powders are used to make a medicinal tea that helps to lower the Hemoglobin A1C and hence an antidiabetic agent. It is also promoted as a weight reduction agent and as anti-inflammation herb.

Salix/Willow (*Salix* sp.): The bark of this tree is the natural source of salicylic acid. **Salicylic acid** and its derivatives (such as Aspirin) are powerful analgesics.

Sanguinarine – (Bloodroot -Sanguinaria canadensis, Mexican prickly poppy-Argemone Mexicana, greater celandine -Chelidonium majus, amur cork tree -Phellodendron amurense). Sanguinarine is the active alkaloid promoted as a treatment or cure for cancer, but the U.S. Food and Drug Administration warns that products having bloodroot, or other sanguinarine-based plants, have no proven anti-cancer effects.

Saponins: These are soap like molecules found in the plant families Caryophyllaceae, Sapindaceae, Aceraceae, Hippocastanaceae, Cucurbitaceae, Quinoa and some other plants. Saponins are used as fish poisons. In humans they cause intestinal distress and hence all grains such as quinoa and some legumes should be washed so that all foaming saponins are washed away before consumption after cooking. Saponins are also being investigated as adjuvants for vaccines. The saponin Diosgenin from Dioscorea spp. Is used in the semi-synthetic or biosynthetic production of progesterone and corticosteroids.

Saw Palmetto (*Serenoa repens*): Saw palmetto has been used in men for urination problems such as decreased urine flow, nighttime/frequent urination due to an enlarged prostate gland -benign prostatic hyperplasia- BPH. This palm tree grows in the Caribbean islands, Florida coast and many other coastal areas. Side effects include nausea and vomiting, interference with anti-blood clotting agents, bruising and allergies.

Schizandra (*Schisandra chinensis*): The berries of this plant are used in various Chinese, Korean and Japanese formulations for supporting general health. It is one of the 50 most important herbs in Chinese medicine.

Scopolamine/Hyoscine (– Henbane – *Hyoscyamus niger*, Jimson weed -*Datura stramonium*, Angel's trumpet -*Brugmansia* spp). The alkaloid Scopolamine is a medication used to treat motion sickness and postoperative nausea and vomiting. It is available as trans dermal patches. Side effects include sleepiness, blurred vision, dilated pupils, and dry mouth.

Sea Wormwood (*Artemesia brevifolia*): The plant is rich in santonin which is a powerful antihelminth (Tape worm) and Acaricide (Round worm). Tea infusions of leaves are used.

Seers Sage (*Salvia Divinorum*) is a plant with psychoactive properties that is native to Mexico and Central and South America. Crushed leaves, tinctures, smoke, and chewing are the common methods of ingestion. The active diterpenoid is the most potent natural hallucinogen (Synthetic Lysergic acid is even more potent) known. In small amounts, it can control intestinal motility and hence diarrhea.

Senna/Tinnevelly Senna/Egyptian Senna/East Indian Senna (*Senna alexandrina/Cassia officianalis*): The dried powdered leaves packed in capsules or tablets or drunk as a tea/decoction in water induces bowel movements and acts a laxative. The side effects of prolonged use are: severe stomach pain, severe diarrhea, watery diarrhea, weight loss, worsening constipation after you stop taking senna, enlargement of your fingers and toes, low potassium, confusion, uneven heart rate, extreme thirst, increased urination, leg discomfort, muscle weakness or limp feeling, or nausea, upper stomach pain, itching, loss of appetite, dark urine, clay-colored stools, jaundice (yellowing of the skin or eyes).

Senna-aavaram/Aavarai (*Cassia auriculata*): The leaves have laxative properties. It is also reported to control diabetes. The dried flowers and flower buds are used to make tea for diabetes patients.

Shandilay (*Leonotis nepetifolia*): Both leaves and flowers have Leonurine which has sedative effects. The plants are commonly used in the Caribbean islands.

Shikakai (*Acacia concinna*): This is a tropical legume shrub. The fruit pods and seeds are ground into a paste and applied to hair. It acts as a natural pH-controlled shampoo. It has some antifungal properties also. It is a good organic shampoo replacement for petrochemical based shampoos.

Silybin also known as silibinin – Milk thistle (*Silybum marianum*): Silibin is marketed in Europe as a drug for treatment of liver cirrhosis. It is also being investigated for removal of iron toxicity in patients who have received multiple iron and blood transfusions. Silibin also has chemoprotective effects in controlling UV induced photo carcinogenesis.

Sinus infections, inflammation and blockage: Extracts of leaf of Mullein (*Verbascum thapsus*), Chamomile tea, Comfrey (*Symphytum uplandica or S. officinale*), Marshmallow (*Althaea officinalis*), extracts and tincture of all above ground parts of Eyebright (*Euphrasia rostkoviana/E. officinalis*), Fennel seed oil (Foeniculum foenum-grecum), Echinacea (*Echinacea sp.*), Eucalyptus oil (*Eucalyptus sp.*), Menthol leaf and essential oil (*Mentha piperata*), Oil of Wintergreen (*Gaultheria sp.*), extracts of Calendula flowers and leaves (Calendula officianalis), root extracts of Astragalus (*Astragalus membranaceus/Astragalus propinquus*), and root extracts Horseradish (*Armoracia rusticana, syn. Cochlearia armoracia*), are all being recommended as herbals for sinus relief. In addition, extracts from the following plants described in Chapter 10 are also used in preparations for treating nasopharyngeal health conditions. They are Mexican mint/Indian Borage/Panikoorkka/Karpooravalli *(Plectranthus amboinicuse)*, Mullein/ Aarons rod (*Verbascum Thapsus*), Opium poppy (*Papaver somniferum*), Thyme (*Thymus vulgaris*), Turkey Bush (*Eremophila gilesii*), Turmeric and white turmeric (*Curcuma longa* and *C. zedoaria*), Long pepper (*Piper longum*); Bibhitaki (*Terminalia bellirica*), Chebulic myrobalan (*Terminalia chebula*), Indian Sarsaparilla (*Hemidesmus indicus, Syn Periploca indica*

Linn), Kizha Nelli (*Phyllanthus niruri*), Black pepper (*Piper nigrum*), Ginger (*Zingiber officianalis*), Holy Basil (*Ocimum sanctum*), Agnidamani (*Solanum Trilobatum*), Lily bulbs (Bai He, Lilium sp), Bitter Kola (*Garcinia kola*), Licorice plant (*Helichrysum petiolare*), African Teak (*Milicia excelsa*), Orange climber (Toddalia asiatica), Paper bark acacia (*Vachellia sieberiana*), Ipecac (*Cephaelis ipecacuanha*), Putty root (*Aplectrum hymale*); Indian tobacco (Lobelia inflate), Maiden hair fern (*Adiantum Capillus-veneris*); Dogwood, (*Cornus florida, syn. Benthamidia florida*), Willow (*Salix sp*), Caraway/Kala Jeera/Perum Jeerakam (*Carum carvi*), Musk Okra (*Hibiscus abelmoschus*, Syn *A.moschatus*), Ajwain/Carum (*Trachyspermum ammi, T. copticum*) and Anise: (*Pimpinella anisum*).

Skullcap (*Scutellaria lateriflora*): The whole plant is rich in several flavones. Plant extracts induce gentle sleep.

Slippery Elm Bark (*Ulmus rubra/U. fulva*): The bark extracts function as demulcents which coat the linings of the intestine and hence reduce inflammation of the bowels. It is said to be an herbal treatment choice for relief of irritable bowel syndrome.

Small Caltrops/Tribulus/devils weed/Devils thorn./Gneriginil/ (*Tribulus terrestris*): Extracts of this plant are reported to boost testosterone levels. It is used as a supplement by body builders and athletes.

Snake gourd (*Trichosanthes cucumerina and T. kirilowii*): The fruit (vegetable) is a part of many vegetable dishes in Africa and Asia. It has abortifacient properties.

Soapnut (*Sapindus saponaria*): The fruit pulp and dried fruit pulp from this tree are good organic soaps. Soapnut powder foams when added to water and can be used as skin cleansers. It may have anti-microbial properties.

Soybean (*Glycine max*): Soybean seed extracts have phytohormones and help in control of hot flashes. Soybean oil is rich in Omega 3 and Omega 6 polyunsaturated fats and hence good for cardio protection. It is a great source of vegetable protein.

Spearmint (*Mentha spicata*): Spearmint is reported to relieve symptoms of digestive problems, by relaxing the stomach muscles, reducing symptoms of nausea, and other digestive problems. It also has anti-fungal properties.

Sterols: Sterols or steroid alcohols are a subgroup of the steroids occurring naturally in plants, animals, and fungi. There are some sterols that have medicinal value. These include beta sitosterol, sigmasterols, campesterol and tocopherols. Beta-Sitosterol found in vegetables, nuts, fruits, and seeds has the potential to reduce benign prostatic hyperplasia (BPH). Likewise, Campesterol found in Banana, pomegranate, pepper, coffee, grapefruit, cucumber, dandelion, onion, oat, potato, lemon grass and rapeseed are an anti-LDL cholesterol agent. Other phytosterols like Stigmasterol (soybean, rapeseed) also reduce LDL cholesterol. Tocopherols (Vitamin E compounds) in Olive oil, sunflower oil, soybean, corn oil were prescribed as agents for protecting heart. But, US FDA, US NIH and the European Food Safety Authority have rejected claims that vitamin E has any positive effect on heart.

Sticky Nightshade, Wild Tomato (*Solanum sisymbriifolium*): The roots of this perennial herb, have been used as a traditional medicine possessing diuretic and antihypertensive properties in Paraguay.

Stinging Nettle (*Urtica dioca*): This prickly plant has also been used to reduce nighttime bathroom trips.

Strychnine – (Strychnine tree -Strychnos nux-vomica): This is a toxic, colorless, bitter, crystalline alkaloid used as a pesticide, particularly for killing small vertebrates including fish. In humans, it causes poisoning which results in muscular convulsions and eventually death.

Sugar destroyer/Charkkarai Kolli/Gurmar/Madhunashini (*Gymnema sylvestre*): Chewing on the leaves of this plant can temporarily interfere with the ability to taste sweetness and hence the desire to eat sweet foods gets reduced. It may help stimulate insulin secretion and the regeneration of pancreas islet cells — both of which can help lower blood sugar.

Sulforaphane/Glucoraphanin (Broccoli): Sulforaphane are glucosinolates considered to be chemo prevention agents. They have other effects including lowering of blood lipids.

Swamp Magnolia (*Magnolia virginiana*): The leaves or bark powders are inhaled as a mild hallucinogen.

Tamarind (*Tamarindus indica; T. occidentalis T. officinalis*): The fruit and fruit extracts commonly used in cooking aids smooth bowel movements since it is a mild laxative. The leaves are added to hot water for bathing since it is believed to control body pains.

Tangeritin (Citrus-tangerine and various citrus peels): Tangeritin is found in in the peel of tangerines and other mandarin oranges. It is found in smaller amounts in the peels of some other citrus fruits. According to some in vitro studies, Tangeritin may help prevent certain types of cancer

Tea (*Camellia/Thea sinensis*): Green tea is said to lower blood pressure and reduce stroke, lower cholesterol.

Terpenoids: Terpenoids contribute to the aroma of many spice plants as well as eucalyptus, citral, menthol, camphor, salvinorin A cannabinoids found in cannabis, ginkgolide and bilobalide found in Ginkgo biloba, and the curcuminoids found in turmeric and mustard seed.

Thumbai poo/Dronapushpa//Goma Madhupathi (*Leucas aspera*): This member of the Lamiaceae/Labiatae (mint) family has white flower and elongated leaves. Both the flowers and leaves have medicinal value. The flowers are boiled in sesame oil to make an herbal oil that cleanses the skin, controls itching and allergies. The leaf juice removes intestinal worms.

Thunder Vine/lei gong teng/, Radix (*Tripterygii wilfordii*): Extracts of this plant have been used in traditional Chinese medicine for control of Rheumatoid arthritis (RA). However, it has severe side effects such as gastrointestinal distress, skin rashes, cardiovascular effects, and other side effects. In view of this use of this herbal for treating RA is not recommended.

Thyme (*Thymus vulgaris*): Thyme leaf essential oil is used in disinfectants as an anti-bacterial agent.

Triphala (mixture of three fruit extracts from Amala/Nellikkai (*Emblica officinalis*), Bibhitaki (*Terminalia bellirica*), and Haritaki (*Terminalia chebula*): This polyherbal Ayurvedic mix prepared as a dry powder or encapsulated in capsules or tablets is a very effective herbal medicine for control of constipation by helping to have regular bowel movements. The most powerful benefits of Triphala include improving digestion, reducing signs of aging, detoxifying the body, aiding weight loss, strengthening the immune system, managing diabetic symptoms, and perfecting nutrient uptake efficiency. As noted previously this poly herbal preparation helps regular bowel movements and has many other health benefits. Overall, this combination of herbals help supports good health.

Turkish Mint (*Lagochilus inebrians*): The active compound Lagochilin has sedative, hypotensive and hemostatic effects. Leaf powders are steeped in hot water and drunk as a relaxing tea by Turkish, Tajik and Uzbek tribes.

Turmeric (*Curcuma longa*): Curcuma longa holds phytochemicals that has anti-inflammatory, anti-microbial, anti-oxidative and many other properties that help to reduce age-related diseases. Powder from the rhizome and other extracts from the turmeric rhizome are popular Ayurveda remedies for treating skin and scalp infections.

Valerian (*Valeriana officinalis*): Valerian root extracts are used to induce sleep. It is a common ingredient along with Melatonin in many sleep pills.

Veldt grape/Adamant Creeper/Asthisamdhani/Pirandai (*Cissus quadrangularis*): This creeping vine belonging to the grape family is listed as a medicinal plant in Africa, South East Asia. It is considered to benefit treatment of Diabetes, high cholesterol, and weight control. It is available as powders, capsules, and syrups.

Welwitsch (*Strophanthus welwitschii*): Root powder in oil is used in Africa for treating scabies.

Vinca: Look under Madagascar Periwinkle and Catharanthus roseus.

Vincristine – (Madagascar periwinkle-*Catharanthus roseus*). Vincristine, also known as leurocristine is a chemotherapy medication used to treat acute lymphocytic leukemia, acute myeloid leukemia, Hodgkin's disease, neuroblastoma, and small cell lung cancer, among others.

Vetiver/Khas/Ramacham Ver (*Vettiveria zizanioides/Chrysopogon zizanioides*): This member of the Poaceae grass family has very aromatic roots containing an essential oil. Both the roots and essential oil are used in Aromatherapy. The roots have a pleasant smell and is put in drinking water containers as a water purifier and to keep the water having a pleasant smell. Medicinally, the oil is used to control inflammations, improve peripheral blood circulation, and improve skin health. Infusions of Vetiver have a calming effect and reduces stress.

Wheat bran (*Triticum aestivum*): Dietary wheat bran intake of 3 to 6 g/day modestly reduces systolic and diastolic BP.

White Magnolia (*Galbulimima belgraveana*): This plant is rich in alkaloids. These include Himbacine, himbeline, himandravine, himgravine, himbosine, himandridine, himandrine, himgaline, himbadine and himgrine.

White bean extract (*Phaseolus vulgaris*): These extracts have an alpha amylase blocker and prevents uptake of carbohydrates. Thus, it is believed to help in weight reduction and lower cholesterol.

Wild Castor/Danti (*Baliospermum montanum/Jatropha montana,*): Root extracts are used as laxatives. Paste from roots is applied on hemorrhoids for shrinking them and on inflamed areas for control of inflammations.

Witch Hazel (*Hamamelis* sp.), Six species namely, *H. Mexicana, H. Ovalis, H. Virginiana, H. vernalis, H. Japonica* and *H. Mollis* are collectively referred as of witch hazels: Witch hazel bark and leaf extracts have anti-inflammatory properties and are also used to tighten the skin and treat acne. Witch hazel extracts are also added to some skin lotion and after shave lotions to keep skin moist and prevent minor bacterial infections.

Woolly morning glory (*Rivea corymbosa/Argyreia nervosa*): The seeds have D-lysergic acid amide, lysergol, and turbicoryn, lysergic acid alkaloids which are all psychoactive.

Wormseed (*Chenopodium ambrosioides anthelminticum/Dysphania ambrosioides*): Chenopodium seed powder and oil are well known herbal vermifuges and are known to expel tapeworms and other intestinal parasites.

Yellow wood (*Bleekeria vitensis/Ochrosia vitensis/O. moorei and Excavatia coccinea*): Reported to reduce the risk of some cancers such as colorectal, breast, liver, lung, prostate, skin, stomach, and bladder cancer.

Yohimbe (*Pausinystalia johimbe*): Bark extracts of Yohime are promoted as medicinal for erectile dysfunction. The prescription drug Yohimbine hydrochloride is derived from this plant. Neither Yohimbe extracts nor the alkaloid Yohimbine support the claims made by marketing outfits. It has a lot of side effects and people with kidney issues, those on blood thinners should not take this preparation.

Yopo (*Anadenanthera peregrina*): Flower extracts are known to cause delirious flight of thoughts, often lasting many days. It has been used in healing ceremonies and rituals for thousands of years in South America.

References

Ajesh TP, Krishnaraj MV, Prabu M and Kumuthakalavalli R. Herbal Abortifacients Used by Mannan Tribes of Kerala, India. International Journal of PharmTech Research 2012; 4 (3): 1015-

Jose A Fernández-Albarral, Rosa de Hoz, Ana I Ramírez, Inés López-Cuenca, Elena Salobrar-García, María D Pinazo-Durán, José M Ramírez, Juan J Salazar (2020): Beneficial effects of saffron (*Crocus sativus* L.) in ocular pathologies, particularly neurodegenerative retinal diseases, Review Neural Regen Res. 2020 Aug;15(8): 1408-1416. doi: 10.4103/1673-5374.274325.PMID: 31997799 PMCID: PMC7059587 DOI: 10.4103/1673-5374.274325

Hamidpour R, Hamidpour S, Hamidpour M, Shahlari M. Camphor (*Cinnamomum camphora*), a traditional remedy with the history of treating several diseases. International Journal of Case Reports and Images 2013;4(2): 86-89.

Matheus Alves de Lima Mota, José Saul Peixoto Landim, Thiago Sousa Silva Targino, Silvia Fernandes Ribeiro da Silva, Sônia Leite da Silva, Márcio

Roberto Pinho Pereira (2015): Evaluation of the anti-inflammatory and analgesic effects of green tea (Camellia sinensis) in mice. Acta Cir Bras. 2015 Apr;30(4): 242-6. doi: 10.1590/S0102-865020150040000002. Epub 2015 Apr 1

Tran TA, Ho MT, Song YW, et al. Camphor induces proliferative and anti-senescence activities in human primary dermal fibroblasts and inhibits. UV-induced wrinkle formation in mouse skin. Phytotherapy Research 2015 Dec;29(12): 1917-1925.

Web References

https://www.cancer.gov/

www.ipni.org

https://www.mayoclinic.org/

https://www.mskcc.org/cancer-care/diagnosis-treatment/symptom-management/integrative-medicine/herbs/search?letter=A

https://www.ncbi.nlm.nih.gov/books/NBK92755/

https://www.artofliving.org/in-en/ayurveda/ayurvedic-remedies/home-remedies-for-diabetes

https://www.nccih.nih.gov/health/bilberry

http://pennstatehershey.adam.com/content.aspx?productid=107&pid=33&gid=000225#Plant%20Description

https://www.mayoclinic.org/drugs-supplements

https://pubmed.ncbi.nlm.nih.gov/322905cancer and promote prostate heaslth. 7/

https://pubmed.ncbi.nlm.nih.gov/30813433/

https://www.exoticindiaart.com/book/details/database-on-medicinal-plants-used-in-ayurveda-seven-volumes-IDG237/

https://en.wikipedia.org/wiki/Chinese_herbology#Chinese_ginseng

http://tropical.theferns.info

https://pfaf.org/user/Default.aspx

https://images.search.yahoo.com/search/images;_ylt=A0geK.M39T5fPBUA6ydXNyoA;_ylu=X3oDMTEydGlpaGkwBGNvbG8DYmYxBHBvcwMxBHZ0aWQDQzAxNjRfMQRzZWMDc2M-?p=Medicinal+plants+PROTA&fr=mcafee

https://uses.plantnet-roject.org/en/PROTA,_Introduction_to_Medicinal_plants

http://www.missouribotanicalgarden.org/plantfinder/plantfindersearch.aspx

https://www.researchgate.net/publication/238506043_Phytochemistry_traditional_uses_and_pharmacology_of_Eugenia_jambolana_Lam_Black_plum_A_review

Glossary

Plant Biology

Algae: These are Unicellular, colonial, filamentous or leaf-like, photosynthetic organisms without conducting elements.

Alternate: Leaves or flowers borne singly at different levels along a stem.

Angiosperms: These plants produce seeds with a covering. They are divided into Basal Angiosperms, Monocots and Eudicots.

Annual/Biennial/perennial: Annual plants are those that complete their life cycle and dies within one year (Annual). Biennial plants complete life cycle in two years. Usually, they grow vegetative during first year without flowering and flower and set seeds in second year. Perennial plants live for an indefinite period of life growing vegetative and setting flowers/fruits and seeds each year.

Anther: Pollen-bearing part of the stamen. Anthers have 2-4 hollow lobes having pollen grains.

Anthesis: Period during which pollen is presented and/or the stigma is receptive.

Apical meristem: the growing point at the apex of the plant.

Aquatic: Plants whose natural habitat is water.

Arborescent: Tree-like growth or general appearance.

Arboretum: A taxonomically arranged collection of trees.

Arid/semi-arid: Regions of dry weather with low rainfall under near desert conditions.

Bark: The protective external layer of tissue consisting of the food-conducting phloem elements on the stems and roots of trees and shrubs.

Binomial Nomenclature: The system of naming biological organisms in Latin developed by Swedish scientist Carl Linnaeus. Here, the scientific name consists of two names, the first name being the generic name and the second the species name. The generic and species names are followed by the name of the variety and or cultivar followed by a reference to the name of the person who named the organism first.

Bisexual: are flowers bearing both male and female reproductive organs.

Bole: The part of the stem below the lowest branch.

Bryophyte: These are a group of non-vascular plants classified as mosses, hornworts, and liverworts.

Bulb: Underground fleshy organ consisting of a stem and leaf bases. Examples: Onion bulbs, tulip bulbs.

Calyx: All the outer components of a flower. They are usually green becoming yellow as flower ages.

Carpel: The basic female reproductive organ in angiosperms.

Chimera: an individual composed of two or more genetically different tissues.

Chlorophyll/Anthocyanin: Chlorophylls are green pigments in chloroplasts, essential for photosynthesis. Anthocyanins are pigments in flowers, ageing leaves and in some other organs such as roots of carrots, beets, and turnips.

Clade: The taxonomic hierarchy of a group of organisms classified together based on homologous features traced to a common ancestor.

Compound leaves: Leaves composed of several leaflets.

Corm: Fleshy, swollen underground tuberous stem storing food reserves.

Cotyledon: The first leaves of an embryo

Common& popular names of plants: Names of plants in English or local ethnic languages.

Community: refers to plants that characteristically occur together in nature.

Cultivar: This refers to cultivated plants that are differentiated by one or more characters.

Decumbent: Branches growing horizontally on the ground but turned up at the ends.

Dicotyledonae/dicots: These are flowering plants with two cotyledons or seed leaves and leaves with netted veins and pollen with three sacks.

Dioecious: When male and female reproductive structures develop on separate male and female plants.

Embryo: young plant contained by a seed.

Endosperm: Nutritive tissue surrounding the embryo of the seed.

Endemic: having a natural distribution restricted to a geographic region.

Epiphytes: are plants living on other plants but not parasitic on them. Many orchids are epiphytes.

Eudicots: The original Dicotyledonae renamed to reflect their morphology that consists of two cotyledons, netted leaf veins, presence of taproots, xylem and phloem arranged in distinct bundles in rings.

Evergreen: Plants that produce leaves all the year round.

Exocarp: the outer layer of the pericarp, often the skin of fleshy fruits.

Exotic: plants that are not native, introduced from another region or country.

Family: A group of plants with many similar qualities and include many related genera and species.

Female reproductive system: ovary and ovules.

Fiber: Fibers may be soluble in water or may be insoluble as in cellulose fibers. Both types of fibers are important for good gastrointestinal and cardiovascular health.

Filament: The male reproductive system known as the stamen consists of filaments bearing the anthers.

Fruit: seed-bearing structure in angiosperms formed from the ovary, and sometimes associated floral parts, after flowering.

Fungi: These are commonly known as molds. They are not photosynthetic and live as saprophytes or parasites.

Genus: (plural, genera): Genus is the principal category of taxa intermediate in rank between family and species in the nomenclatural hierarchy.

Genotype/Phenotype: the genetic make-up of an individual is called the genotype and the physical appearance of the plants (morphology) is called phenotype. 2n=14 is genotype of wheat, 2n=46 is the genotype of humans. The physical description of the plant is called the phenotype.

Glutinous: Sticky seeds having gluten. Rice seeds may be glutinous (Japanese varieties) or not.

Group: refers to an assembly of two or more cultivars within a species or hybrid. For example, bean group consists of *Phaseolus vulgaris var cranberry*, *P. vulgaris var pinto* etc.

Gymnosperms: Seed-bearing plants with naked ovules borne on the surface of sporophylls. Examples are conifers, Ginkgo, Gnetum and cycads.

Habit: The general external appearance of a plant, including size, shape, texture, and orientation.

Habitat: The place where a plant lives, the environmental conditions of its home.

Halophyte: A plant adapted to living in highly saline habitats like salt marshes, ocean beach shorelines.

Haploid/Diploid/Tetraploid/polyploids: Haploid: Cells have one copy of each chromosome. For example, the haploid chromosome number of Humans is n=23, in wheat it is n=7etc. This is the case in the gametes of all organisms irrespective of sex. In diploids represented by the human body (somatic) it is 2n=46, in wheat it is, 2n=14. Higher than diploids in humans is not viable. However, in plants, triploids (3n), tetraploid (4 n) and polyploids etc are common. Polyploidy: refers to any organism with more than two of the basic sets of chromosomes in the nucleus. Various combinations of words or numbers with '-ploid' show the number of haploid sets of chromosomes, e.g. triploid = three sets, tetraploid = four sets, pentaploid = five sets, hexaploidy = six sets, etc.

Herb/Shrub/Tree/Vine/Succulents: Herbs are Vascular plants that do not develop a woody stem, for example, Corn, rice, chili pepper, tomato, eggplant, beans. Shrub is a woody perennial plant without a single main trunk, branching freely, and smaller than a tree. Trees are woody plants, usually with a single distinct trunk and several meters/feet tall. Vine is a plant that twines around a substrate like a pole, e.g. beans. Succulents are plants with thick leaves having fluids such as Cactus.

Herbaceous: Herb-like.

Herbarium: is a collection of preserved, usually dried, plant material. In addition, a building in which such collections are stored.

Hip: This is a term referring to the fruit of a Rose.

Hybrid: The progeny resulting within and between the crossings of two different species, varieties, cultivars etc of plants are called hybrids.

Kingdom: This term refers to Category of organisms at the highest level in taxonomy for example, Kingdom: Plantae.

Inflorescence: several flowers closely grouped together to form an efficient structured unit.

Latex: is a milky fluid that exudes from plants such as rubber, figs, dandelions, and milkweed.

Lithophyte: Plants that grow in rocky regions.

Magnoliids: These are angiosperms characterized by flowers whose components are arranged in groups of three, pollen with one pore with branched leaf veins: Examples are Avocado, magnolias, black pepper.

Mangrove plants: are shrubs or small trees growing in salt or brackish water, usually characterized by pneumatophores that are breathing roots. The Everglades in Florida, USA, Sundarbans, and Pichavaram in India are examples of Mangrove plants.

Meristems: are actively dividing regions in the plant that manage overall growth of plant. Terminal meristems found at the apex of stem and roots of plants manage growth in height. The axillary meristems found in axils of eaves give rise to branches and intercalary meristems give rise to growth in width. Leaf meristems found in the periphery of leaf blades manage growth of leaves, and floral meristems give rise to flowers. Meristems have cells that have properties like stem cells in humans.

Monoecious: All flowers are bisexual on same plant or male and female unisexual flowers on the same plant or of an inflorescence, which has unisexual flowers of both sexes.

Multiple fruit: are clusters of fruits produced from more than one flower appearing as a single fruit, often on a swollen axis.

Mutation: This refers to abrupt and inexplicable variation from the norm, such as the doubleness in flowers, changes in color, or habit of growth. Mutations can result sporadically due to genetic mistakes during DNA replication or because of exposure to irradiation or chemicals.

Monocotyledonae/Monocot: This refers to angiosperms that have only one cotyledon in the seed. All grasses, all cereals, banana, bamboo, coconut, are examples.

Nectary: A specialized gland that secretes nectar, honey.

Nut: is a hard, dry, indehiscent fruit, having only one seed.

Ovary: is the basal part of a carpel or group of fused carpels, enclosing the ovule(s).

Ovule: The seed before fertilization

Order: In taxonomic hierarchy, this refers to a group below that of the Division (Division) or clade and just above the family.

Parthenocarpy: refers to the production of fruit without pollination. Thus, these fruits are seedless. Parthenocarpy can be produced naturally or through chemical stimulation by plant hormones like gibberellin acid.

Perennial: plants whose life span extends over several years.

Population: All individuals, or of one or more species within a prescribed area.

Procumbent: Spreading along the ground but not rooting at the nodes

Pteridophytes: This Includes all ferns but not gymnosperms or angiosperms.

Pericarp: is the outer layer of the wall of a fruit namely the 'skin'.

Petals: Parts of a flower with colored petals. A group of petals is called a corolla.

Phloem: is a specialized conducting tissue in vascular plants that transports sucrose and other metabolites from the leaves to other plant organs.

Pistil: A single carpel when the carpels are free or a group of carpels when the carpels are united by the ovary.

Pod: is the fruit of a leguminous plant, a dry fruit of a single carpel, splitting along two sutures.

Pollen: are powdery mass shed from anthers (of angiosperms) or microsporangia (of gymnosperms). They contain the male sperms.

Pollination: is the transfer of pollen from the male organ (anther) to the receptive region of a female organ (stigma).

Saprophyte/epiphyte/Lithophyte/parasite: Organisms deriving nourishment from decaying organic matter are Saprophytes; Plants living on the surface of another plant without being a parasite are Epiphytes. Plants that live in a rocky substrate are Lithophytes. Parasites are plants that derive all nourishment from another. Parasites may be facultative in which case they can either parasitize another plant or live like a saprophyte or they may be obligate in which case they are entirely dependent on the host for all their growth requirements.

Rhizome: Perennial underground horizontal stem.

Seed: A fertilized ovule made of a protective seed coat enclosing an embryo and food reserves.

Sepals: are the outermost parts of a flower. These are usually light/dark green. Collectively sepals are called calyx.

Simple leaf: Undivided leaf.

Stamen: male organ of a flower in angiosperms, consisting of a stalk (filament) and a pollen bearing part (anther). Staminate flowers are male flowers while pistillate flowers are female flowers.

Staminode: is a sterile stamen, often rudimentary, sometimes petal-like.

Stigma: The pollen-receptive surface of a carpel at the summit of the style.

Style: Elongated part of a carpel, or group of fused carpels topped by the stigma.

Tendril: a slender organ used by climbing plants to cling to an object.

Tuber: an underground storage organ formed by the swelling of an underground stem which produces buds and stores food, forming a seasonal perennating organ such as potato or yam. Swollen roots are also referred to as tubers.

Unisexual: flowers of one sex, bearing only male or only female reproductive organs.

Zygote: is a fertilized cell, the diploid product of fusion of two haploid gametes.

Species: refers to populations of individuals that share common features and/or ancestry. Examples are *Triticum aestivum*, *Triticum vulgare* (wheats of two different species).

sp.: abbreviation of species often used when the genus but not the species is clear.

spp.: plural of sp.

Taxonomy: Classification of all living organisms.

Temperate: Geographical areas with lower/cooler temperatures.

Terrestrial plants: Land plants.

Tropical and Sub-Tropical: Geographical areas with warmer temperatures usually near the equator and between the tropic of cancer in the North and tropic of Capricorn towards the south.

Type species: This refers to a model kept in an herbarium for naming and describing an organism.

Var: (variety in common usage, abbreviated as var: In the binomial system, a variety ref **Vascular tissue:** The conducting xylem and phloem elements.

Xerophyte: Plants with succulent stems living in desert-like conditions, example: Cactus.

Xylem: a specialized water-chemical conducting tissue in vascular plants.

Plant Chemistry

Alkaloids: These are a class of naturally occurring organic nitrogen-containing bases derived from plants. They include cocaine, nicotine, strychnine, caffeine, quinine, ephedrine, morphine, Pilocarpine, atropine, methamphetamine, mescaline, ephedrine, and tryptamine. They have a bitter taste.

Amenorrhea: Absence of menstruation in timely fashion.

Anthocyanin: These are water-soluble plant pigments ranging in color such as blue, red, purple found in flower petals, fruit, and vegetable skins.

Arteriosclerosis: Hardening of blood vessals.

Carbuncle: Large boil on skin with discharges usually caused by staphylococcus infections.

Carotenoid: Carotenoids are any of a class of yellow, orange, or red fat-soluble pigments, including carotene, which give color to plant parts such as ripe tomatoes and autumn leaves. They are terpenoids. Carrots are rich in carotene. Carotene is a precursor of Vitamin A.

Chlorophyll: Chlorophylls are green pigments found in the chloroplasts of plants particularly leaves.

Cyano-glycosides and glycosides: Glycosides are compounds formed from a simple sugar and another compound by replacement of a hydroxyl group in the sugar molecule through a glycosidic bond. Cyano glycosides have a nitrile group. The enzyme beta glycosidase reacts with cyano glucosides to release the toxic cyanide. Example: amygdalin in bitter almonds. Apple seeds, peach seeds, flax seeds, tapioca all have cyanoglucosides.

Flavones: A crystalline compound, $C_{15}H_{10}O_2$, that is the parent substance of several yellow plant pigments.

Flavonoids: Flavanoids are a large group of biologically active water-soluble plant chemicals that include pigments ranging in color from yellow to red to blue and occur especially in fruits, vegetables, and herbs.

Glyco-proteins: Glycoproteins are proteins where the amino acids are covalently attached to Oligo saccharides (more than two sugars).

Monounsaturated Oil: These are fatty acids having one double bond: Examples are Olive oil, Sunflower oil, avocado oil.

Omega-6 oil: These are polyunsaturated fatty acids characterized by a double bond six atoms away from the terminal methyl group.

Omega-3 oil: These are polyunsaturated fatty acids characterized by a double bond three atoms away from the terminal methyl group.

Phytochemical: Chemical compounds found in plants.

Phytomedicine: All medicines derived from plants.

Polyunsaturated: These are fatty acids having more than one double bond. Examples, Walnut and canola oils.

Proteins: Proteins are chains of amino acids. Two or more of these amino acid chains are called peptides. If there are two amino acids then it is a dipeptide, three are tripeptide and multiple are Oligo peptides and many are called polypeptides.

Saturated oil: these are fatty acids with no double bonds: Example: coconut oil, Butter.

Terpenes: Terpenes are the naturally occurring combination of carbon and hydrogen found in essential oils. Examples: Cannabinoids in Cannabis sativa.

Terpenoids: Terpenoids are terpenes that have been changed through a drying and curing process (chemical modification), altering the oxygen content of the compound.

Trans-fatty oils: These fatty acids are artificially saturated by hydrogenation. Example: Margarine.

Xanthophyll: Xanthophyll is a carotenoid yellow- brown pigment in ageing leaves.

Human Biology-Medicine and Pharmacology

Acne: Acne rosacea is a skin disorder leading to pimples, redness on nose, cheek, and forehead.

Adenoids: Lump behind mouth which if large can affect breathing and hearing.

Astringent: Astringents are agents that shrink body fluids. Mouth washes, cough syrups witch hazel, rubbing alcohol, tannins, poly phenols, Athletes-foot medications are examples.

Cardiotonic: Cardiotonic is a substance that has a favorable effect upon the action of the heart. Drugs that have beneficial action on the heart are classed as Cardiotonic.

Carminative: Carminatives are agents that are digestive stimulants that help to reduce cramping, expel gas, and reduce flatulence. Examples are ginger, Dill Seeds, Cumin, Anise, peppermint, coriander,

Cholagogue: A Cholagogue is a medicinal agent that promotes the discharge of bile from the liver/gallbladder into the duodenum.

Colic: Pain due to intestinal gas, passing of gall stones, passage of kidney stones or lead poisoning.

Diaphoretic: These agents promote perspiration.

Dropsy: Swelling mostly in ankles and feet but could be anywhere in body due to fluid retention.

Gallstones: These are stones known as cholesterol gallstones and pigment gallstones found in Gall bladder, which can block flow of bile into the bile duct.

Goiter: Enlargement of thyroid glands,

Kidney stones: Stones due to deposition of calcium oxalate or urea crystals in kidney, ureters, or bladder.

Hepato toxic: Refers to anything that poses liver malfunction.

High-density lipoprotein (HDL): HDL proteins are lipoproteins of blood plasma that is composed of a high proportion of protein with little triglyceride, cholesterol, and that is correlated with reduced risk of atherosclerosis: These are also called good cholesterol.

Kidney stones/Nephrolith: Pebble-like Stones found in kidneys usually of Calcium oxalate or Uric acid and sometimes of cysteine and magnesium ammonium phosphate (Struvite).

Lipo proteins: These are proteins linked to fatty acids and are divided into Low density, Intermediate density and High-density lipo proteins.

Low-density lipo protein (LDL): LDL proteins are lipoproteins of blood plasma that is composed of a moderate proportion of protein with little triglyceride and a high proportion of cholesterol and that is associated with increased probability of developing atherosclerosis. These are also called Bad Cholesterol.

Neuritis: Inflammation of nerves usually peripheral blocking sensory perceptions. Diabetes can lead to peripheral neuritis.

Orchitis: Inflammation of the testicles.

Pleurisy: Inflammation of the double layered membrane surrounding the lungs usually due to fluids caused by viral or bacterial infections.

Scabies: Itching caused by burrowing of skin by mites.

Sialagogues: These agents promote salivation (flow of Saliva).

Triglycerides: Triglycerides are esters formed from glycerol and three fatty acid groups. Triglycerides are the main constituents of natural fats and oils, and high concentrations in the blood show an elevated risk of stroke.

Urticaria: Another name for Hives. Symptoms include red itchy raised areas on skin often caused by allergies and sometimes by certain infections.

Vitiligo: Known also as leucoderma. White sopts or patches appear on skin usually near lips and anywhere due to depigmentation. Caused by genetics, autoimmune and environmental factors.

Yeast infections: Caused by the fungus *Candida albicans* which is a normal fungus found in intestine and locally in vagina. Candida grows excessively during antibiotic treatments, steroids, and contraceptives. Vaginal itching is often caused by candida.

www.ingramcontent.com/pod-product-compliance
Lightning Source LLC
Chambersburg PA
CBHW020853180526
45163CB00007B/2493